ELECTRONIC ANALYSIS INSTRUMENTS

ELECTRONIC ANALYSIS INSTRUMENTS

Harry N. Norton

Harry N Norton
ALTADENA, CA
NOV. '94

PRENTICE HALL, Englewood Cliffs, New Jersey 07632

Library of Congress Cataloging-in-Publication Data

Norton, Harry N.
Electronic analysis instruments/Harry N. Norton.
p. cm.
Includes bibliographical references and index.
ISBN 0-13-249426-4
1. Electronic instruments. I. Title.
TK7870.N67 1992
681′.2—dc20 91-14803
CIP

Acquisitions editor: George Kuredjian
Cover design: Karen Marsilio
Copyeditor: Jane Matsinger
Prepress buyer: Mary McCartney
Manufacturing buyer: Susan Brunke

The publisher offers discounts on this book when ordered in bulk quantities. For more information, write:

Special Sales/Professional Marketing
Prentice Hall
Professional & Technical Reference Division
Englewood Cliffs, New Jersey 07632

Printed in the United States of America
10 9 8 7 6 5 4 3 2 1

ISBN 0-13-249426-4

Prentice-Hall International (UK) Limited, *London*
Prentice-Hall of Australia Pty. Limited, *Sydney*
Prentice-Hall Canada Inc., *Toronto*
Prentice-Hall Hispanoamericana, S.A., *Mexico*
Prentice-Hall of India Private Limited, *New Delhi*
Prentice-Hall of Japan, Inc., *Tokyo*
Simon & Schuster Asia Pte. Ltd., *Singapore*
Editora Prentice-Hall do Brasil, Ltda., *Rio de Janeiro*

CONTENTS

PREFACE

The intent of this book is to present a brief but comprehensive survey of the design and operation of all types of electronic instruments used for analysis. For illustrative purposes, many typical instrument designs that are commercially available are shown and explained. Introductory material covers the basics of properties that are subject to analysis, as well as general fundamentals of sensors, instrument systems, and data systems; a check-list of instrument selection criteria is included. A brief bibliography, at the end of the book, lists books from which additional information can be extracted; however, it excludes the numerous conference papers and articles that have appeared in trade and scientific journals.

This book should prove very useful as a reference book in trade schools, colleges, and universities. More importantly, it should be particularly helpful to technicians, engineers, and managers that need to be familiar with modern electronic analysis instruments as part of their jobs. Since there are many more instrument users than manufacturers, this book is aimed primarily at the user community. However, some categories of personnel at instrument manufacturers' facilities can benefit from the book, such as, to some extent, instrument designers and, to a larger extent, entry-level marketing personnel.

"Entry-level," in this context, is not limited to recent graduates with little or no work experience; it can equally well be applied to those with years of experience that, however, did not include a familiarity with modern electronic instruments of the types described in this book.

Electronic analyzers have come a long way from complex laboratory equipment that required very high levels of skill and knowledge to operate it properly. Modern versions of such instruments are now increasingly used on-line, for example, in process control applications, for either information or closed-loop control purposes, or both. Instruments, or at least their sensing portions, have become much more rugged and reliable, so that they can be installed in remote locations and can operate unattended for long periods of time. With the increasing availability of microprocessors, electronic analyzers have also become significantly more user-friendly. Many designs now include provisions for interfacing them with a PC. Moreover, many analyzers provide interfaces with information networks, so that a large number of them can be read out as well as be controlled from a single operator station; ancillary equipment such as printers, plotters, and modems can easily be added to such networks. Signal processing that results in specific types of information displays is often either built into an instrument system or can be added to a network, and this goes a long way toward simplifying information interpretation.

Electronic analysis instruments are now widely used in production, quality assurance, and research, and these applications are expected to grow continuously. Industries and laboratories have learned that use of these modern instruments results in decreased production costs, quicker and more accurate quality control, and increased research capabilities, together with savings in experiment time and more accurate and reliable results. Additionally, use of many types of analyzers is now mandated by governmental regulations relating to environmental pollution as well as occupational safety and health.

The contents of this book place emphasis on those categories and types of instruments that are produced in relatively large quantities by more than one manufacturer, and are used industrially or commercially in production (which includes process measurement and control), quality control, and research. Many of these instrument types, sometimes in modified versions, are also used in basic (e.g., government-sponsored or academic) research. Although instruments specifically designed to produce images (such as television cameras, multispectral imagers, and radar imagers) are not included, a fair number of instrument categories covered are now also used for remote sensing, sometimes in ground installations, but more often installed on satellites and spacecraft.

Harry N. Norton
Altadena, California

ELECTRONIC ANALYSIS INSTRUMENTS

1

INTRODUCTION TO MEASUREMENT AND ANALYSIS

In this introductory chapter, it is assumed that the reader has a reasonable background in chemistry, including analytical and physical chemistry, but may not be as familiar with (or may not remember as much about) solid and fluid mechanics and thermodynamics. Hence, substantially more emphasis is placed on definitions of terms relating to physical properties than on definitions of terms commonly used in chemistry (e.g., solution, reaction, valence, salt).

The purpose of analysis, in the context of this book, is to make determinations about substances. Such determinations are directed at chemical composition, chemical properties, or physical properties. The instruments described in this book are used primarily for determinations of chemical properties and of composition. Instruments (sensors and transducers) for physical measurements (including mechanical analysis) are described in depth in a companion volume to this book (*Handbook of Transducers,* by H.N. Norton, Prentice-Hall, 1989). These, therefore, are excluded from the scope of this book. However, physical properties of substances are briefly described below, and some of the more specialized types of instruments used to determine certain physical properties are covered in Chapter 7 of this book.

1.1 PHYSICAL PROPERTIES OF SUBSTANCES

Physical properties are categorized as solid-mechanical quantities (generally not pertinent to analysis instruments and excluded here), fluid-mechanical quantities (pressure, flow, moisture and humidity, liquid level, density, viscosity), thermal quantities (temperature, heat flux), and optical quantities (light intensity, color characteristics, index of refraction, turbidity, opacity). Basic definitions related to these quantities follow.

1.1.1 Pressure

Pressure is a multidirectionally uniform type of stress, a force acting on a unit area; it is measured as force per unit area exerted at a given point.

Absolute pressure is measured relative to zero pressure (a "perfect vacuum").

Gage pressure is measured relative to ambient pressure.

Differential pressure is the pressure difference between two points of measurement, measured relative to a reference pressure or a range of reference pressures.

Partial pressure is the pressure exerted by one constituent of a mixture of gases not reacting chemically with each other. According to *Dalton's law,* the total pressure in the mixture equals the sum of the partial pressures.

Static pressure is the pressure of a fluid exerted normal to the surface along which the fluid flows.

Impact pressure is the pressure in a moving fluid exerted parallel to the direction of flow (due to flow velocity).

Stagnation pressure (also called *total pressure*) is the sum of the static pressure and the impact pressure. According to *Bernoulli's equation* for horizontal flow, $p_s = p_0 + \frac{1}{2}\rho V_0^2$, where p_s is the stagnation pressure, p_0 the static pressure, ρ the density of the fluid, and V_0 the flow velocity upstream from the stagnation point. The flow velocity can thus be determined as $V_0 = \sqrt{2/\rho(p_s - p_0)}$.

Head is the height of a liquid column at the base of which a given pressure would be developed, for a liquid having a given *specific weight* (density multiplied by the acceleration due to gravity); the pressure p_L at a depth h below the surface of a liquid is given by $p_L = wh$, where w is the specific weight of the liquid.

1.1.2 Flow

Flow is the motion of a fluid.

Flow rate is the time rate of motion of a fluid, expressed as fluid quantity per unit time; the quantity is fluid mass for *mass flow rate* and fluid volume for *volumetric flow rate.*

Total flow is the flow integrated over a specific time interval.

A *flowmeter* is a flow-rate transducer.

Laminar flow (streamline flow) is the motion of fluid particles along lines that run parallel to the local direction of flow; it can be represented as layers of fluid sliding steadily over one another.

Turbulent flow is the motion of a fluid whose velocity at a point of observation fluctuates with time in a random manner.

The *Reynolds number* is a dimensionless number used to express the fluidity of a fluid in motion; it is given by the relationship $N_R = VD\rho / \mu$, where V is the mean flow velocity, D is the internal diameter of the pipe through which the fluid is flowing, ρ is the density and μ the viscosity of the fluid, and N_R is the Reynolds number. Reynolds numbers below about 2000 indicate laminar flow, whereas those above 4000 generally indicate turbulent flow.

1.1.3 Moisture and Humidity

Humidity is a measure of the water vapor present in a gas.

Absolute humidity is the mass of water vapor present in a unit volume.

Relative humidity is the ratio of the water-vapor pressure actually present to the water-vapor pressure required for saturation, at a given temperature. This ratio is expressed in percent and is temperature dependent; it is the most commonly used quantity in humidity measurement, including in weather reports.

Specific humidity is the ratio of the mass of water vapor in a sample of gas to the total mass of the sample.

The *humidity mixing ratio* is the mass of water vapor per unit mass of the dry constituents.

Moisture is the amount of water (unless another liquid is specified) adsorbed or absorbed by a solid, or in a liquid, or chemically bound to a liquid.

The *dew point* is the temperature at which the saturation water-vapor pressure is equal to the partial pressure of the water vapor in the atmosphere. Any cooling of the atmosphere below the dew point would produce water condensation. The term has also been defined as the temperature at which the actual quantity of water vapor in the atmosphere is sufficient to saturate this atmosphere with water vapor. The relative humidity at the dew point is 100%.

1.1.4 Density and Viscosity

Viscosity is a fluid's resistance to the tendency to flow.

Kinematic viscosity is the ratio of the viscosity of a fluid to its density.

Relative viscosity is the ratio of the viscosity of a liquid to the viscosity of a liquid considered as standard; in the case of a solution, it is the ratio of the viscosity of the solution to the viscosity of the pure solvent in it.

The *Prandtl number* of a fluid is the ratio of its kinematic viscosity to its thermal conductivity.

Density (absolute density) is the ratio of the mass of a homogeneous substance or body to its volume (mass per unit volume).

Relative density is the ratio of a mass of a substance to the mass of a reference substance of the same volume.

Specific gravity is the ratio of the density of a substance at a given temperature to the density of a substance considered as standard; pure distilled water at 4°C or at 15.5°C (60.0°F) has been used as such a standard in reference to liquids and solids. Specific gravity varies with temperature.

1.1.5 Liquid Level

Liquid level is the height of the surface of a liquid or a quasi-liquid (e.g., slurry or a powdered or granular solid) relative to a reference point.

Continuous level sensing refers to stepless sensing over a specified height range.

Point level sensing is a discrete measurement of the presence or absence of liquid at a specified height or location. A given installation may, however, employ two or more point level sensors.

The *interface* between two fluids, sometimes a liquid and a gas, but more often two nonmixing liquids, is another parameter detectable by level sensors.

Pressure methods of level sensing are based on a measurement of *head*, the height of a liquid column at the base of which a pressure is developed.

1.1.6 Temperature and Heat

The *temperature* of a body or substance is its thermal state considered with reference to its power of communicating heat to other bodies. It is a measure of the kinetic energy of the molecules of a substance due to heat agitation; or, looked at another way, it is the potential of heat flow.

Absolute temperature is temperature measured on the thermodynamic scale—temperature measured from *absolute zero* (0 K, −273.15°C).

The *boiling point* is the temperature of equilibrium between the liquid and vapor phases of a substance (at 1 standard atmosphere unless otherwise specified).

The *freezing point* is the temperature of equilibrium between the solid and liquid phases of a substance (at 1 standard atmosphere unless otherwise specified).

The *triple point* is the temperature of equilibrium between the solid, liquid, and vapor phases of a substance.

The *ice point* (273.15 K, 0°C, 32.0°F) is the temperature at which ice is in equilibrium with air-saturated water at a pressure of 1 standard atmosphere.

The *sublimation point* is the temperature at which a substance changes from its solid phase directly to its vapor phase without passing through a liquid phase.

Heat is energy, specifically energy in transfer, due to temperature differences between a system and its surroundings or between two systems, substances, or bodies. It has also been defined as the energy contained in a sample of matter comprising potential energy resulting from interatomic forces as well as kinetic energy associated with random motion of molecules in the sample.

Heat transfer is the transfer of heat energy by one or more of the following methods:

1. *Conduction:* diffusion through solid material or through stagnant fluids (liquids or gases).
2. *Convection:* the movement of a fluid (between two points).
3. *Radiation:* electromagnetic waves.

Heat flux is the amount of heat transferred across a surface of unit area per unit time.

Heat capacity is the quantity of heat required to raise the temperature of a system, substance, or body by one degree of temperature (in a specified manner).

Specific heat is the ratio of the heat capacity of a body to the mass (or to the volume, as specified) of the body. An earlier definition made specific heat the (dimensionless) ratio of the quantity of heat required to raise the temperature of a material (without change of phase) by one degree to the quantity of heat required to raise the temperature of water, having the same mass as the material, by one degree (in temperature), under specified conditions.

Thermal conductivity is the ratio of the time rate of heat flow per unit area to the negative gradient of the temperature per unit thickness in the heat-flow direction.

Thermal diffusivity is the ratio of the thermal conductivity to the product of specific heat and density; the magnitude of thermal diffusivity determines how fast temperature differences that exist in a body or substance will equalize.

Thermal resistance is a measure of a body's ability to prevent heat from flowing through it; it is equal to the difference in temperature between opposite faces of the body divided by the heat-flow rate.

Thermal equilibrium is a condition of a system and its surroundings (or of two or more systems, substances, or bodies) such that no temperature differences exist between them (i.e., no more heat transfer occurs between them).

1.1.7 Light (Visible, Ultraviolet, and Infrared Radiation)

Light is a form of radiant energy; it is electromagnetic radiation in that portion of the spectrum lying between 10 nanometers (nm) and 1 millimeter (mm) (1 mm = 1000 μm = 10^6 nm). This definition of the light spectrum is used throughout this chapter. By strict definition, however, only *visible radiation* (visible light), whose spectrum extends from 380 to 780 nm, can be considered "light." The band of wavelengths between 10 and 380 nm is called *ultraviolet (UV) radiation* (ultraviolet light). The band of wavelengths between 780 and 10^6 nm is called *infrared (IR) radiation* (infrared light). Within the IR band, the portion between 780 nm and 3 μm (3000 nm) is sometimes called *near IR,* and when it is, the band between 3 μm and 1000 μm is called *far IR.*

Most of the IR band overlaps the *heat radiation* band of electromagnetic radiation. Optical quantities can be stated in terms of either *visual* or *nonvisual* (thermally radiative) magnitudes, and optical detectors can be divided into two major groups: *quantum detectors* (photon sensors) and *thermal detectors* (thermal radiation sensors).

Spectral characteristics of light are most commonly shown in terms of wavelength. However, they can also be shown in terms of frequency, wave number, or photon energy. The interrelationship of these quantities, and their relationship to black-body temperature, are illustrated in Figure 1–1. The visible-light spectrum is shown in Figure 1–2.

Frequency of light is related to wavelength by the speed of light in vacuum (2.99 793 × 10^{10} cm/s, normally rounded off to 3 × 10^{10} cm/s):

$$\nu = \frac{3 \times 10^{10}}{\lambda}$$

where ν = frequency of light, Hz
λ = wavelength of light, cm

Wave number is the reciprocal of wavelength (in cm); hence, wave number is expressed in cm^{-1}.

Photon energy is related to the frequency of light by *Planck's constant h* according to the equation

$$\mathscr{E}_p = h\nu$$

where $\mathscr{E}_p$ = photon energy, J (*Note:* 1 electron volt = 1.6022 × 10^{-19} joule)
h = Planck's constant, 6.626 196 = 10^{-34} J·s
ν = frequency, Hz

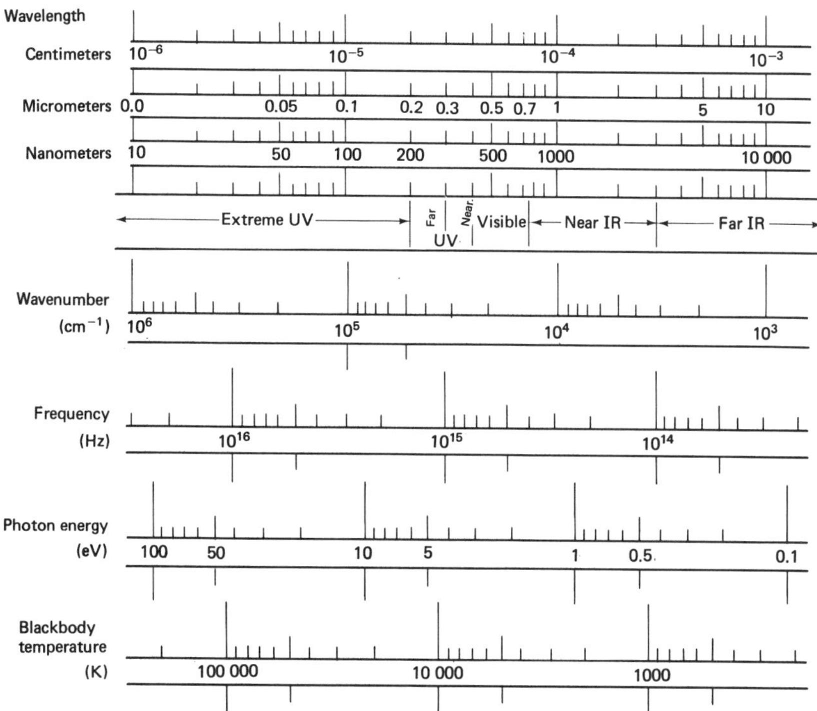

Figure 1-1. The light spectrum.

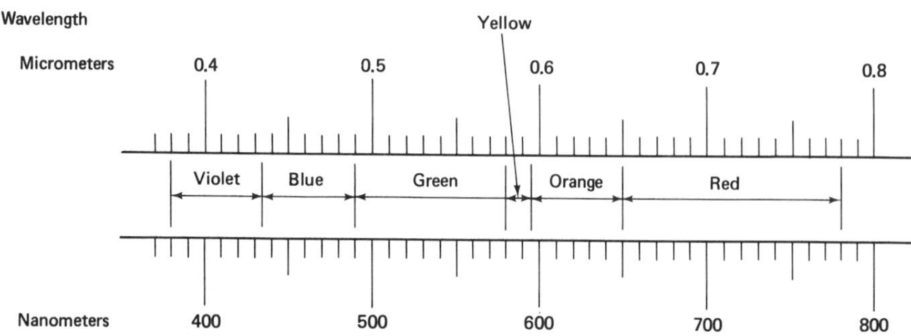

Figure 1-2. The visible-light spectrum.

Black-body temperature, the temperature at which a black body radiates energy such that the radiation has greatest intensity at a certain wavelength, is related to wavelength by *Wien's displacement law,*

$$\lambda_m T = 0.2897 \text{ cm·K (a constant)}$$

where λ_m = wavelength at which radiant energy density is maximum, cm
T = absolute temperature of black-body radiator, K

Luminous flux is the time rate of flow of light; the equivalent nonvisual quantity is *radiant flux.*

Luminous intensity is the luminous flux (emitted by a point source) per unit solid angle; the equivalent nonvisual quantity is *radiant intensity,* the radiant flux per unit solid angle.

Illuminance (illumination) is the luminous flux per unit area of a uniformly illuminated surface on which this flux is incident; the equivalent nonvisual quantity is *radiant flux per unit area.*

Luminance (brightness) is the luminous intensity of a (light-emitting) surface, in a given direction, per unit of (projected) area of the surface, as viewed from the given direction; the equivalent nonvisual quantity is *radiance,* the radiant flux per unit area, per unit solid angle.

Luminosity is the ratio of the luminous flux to the corresponding radiant flux.

A *photon* is the quantum of an electromagnetic field.

1.1.8 Color

Color is a characteristic of visible light associated with its wavelength (see Figure 1.2), especially its dominant wavelength, as well as with purity and with luminance (see definition in Section 1.1.7; visually perceived brightness is approximately proportional to the cube root of luminance).

Purity (colorimetric purity) is the degree to which a primary color is pure (not mixed with the other two primary colors); for any color, it is the ratio of a spectrally homogeneous component to the luminance of the achromatic light with which it must be mixed to match the chromaticity of a sample of light.

Chromaticity is the color quality of light that can be defined by its *chromaticity coordinates,* the ratios of each of the three tristimulus values of a sample to the sum of these values; chromaticity depends on the hue and saturation of a color, not on its luminance; it is also definable by dominant wavelength and purity, taken together.

Hue is the subjectively perceived spectral characteristic of color which determines whether the color is called red, green, greenish blue, reddish purple, and so on; it corresponds to the perceived (psychophysiological) dominant wavelength.

Dominant wavelength is the (single) wavelength of light that matches a color of a sample when combined in suitable proportions with a reference standard light.

Saturation (color saturation) is the degree to which a color differs from a near-gray color of the same perceived hue; it can also be described as the degree to which a color is mixed with white; the more it is mixed with white, the lower is the saturation; for example, low saturation characterizes a "vivid red," whereas high saturation characterizes a "dull red" or "weak red"; saturation has also been equated with and referred to as *chroma.*

Monochromatic light is light in a narrow band of wavelengths (ideally at a single wavelength).

Achromatic color is devoid of any hue; gray is achromatic color.

Color temperature is the blackbody temperature required to produce the same chromaticity as the light being examined.

Tristimulus values are the magnitudes of three different standard stimuli needed to match a given sample of light; the most commonly used standard stimuli are the CIE (International Commission on Illumination or *C*ommission *I*nternationale d'*E*clairage) color mixture functions known as X, Y, and Z, where Y is the *luminosity function* (light-dark), Y–Z corresponds to yellow-blue, and X–Y to red-green; X, Y, and Z can be computed for a sample of light by a color match produced by adjusting the mixture of red, green, and blue lights (the *working primaries*).

Complementary chromaticity is the property of two light samples by which they produce achromatic color (as visually perceived) when combined in suitable proportions.

Chrominance is the difference between any color and a reference color of specified chromaticity and of equal luminance.

Primary colors are colors of constant chromaticity and variable luminance which can produce (or can be used to specify) other colors when mixed in suitable proportions.

1.1.9 Properties Related to Interaction of Light with Substances

Absorption is the process by which a portion of the incident (light) energy is transferred to the substance receiving that energy; this phenomenon is related to characteristic molecular structures of substances.

Absorptance is the ratio of the (luminous or radiant) flux absorbed in a substance to the flux incident on the substance.

Absorbance is the logarithm to the base 10 of the reciprocal of the *tramsmittance.*

Transmittance is the ratio of the radiant power transmitted by a substance to the radiant power incident on the substance.

Reflectance (radiant) is the ratio of the radiant flux reflected by the surface of a substance to the radiant flux incident on that surface.

Refraction is the change in direction of a light beam (or ray of radiation) as it passes from one medium to another medium having a different *refractive index*.

Diffraction is the interference pattern resulting from the uniting, at each point, of rays from different points around an opaque object, or of rays through different parts of an opening.

Interference is the variation, due to the superposition of two or more waves, in wave amplitude, with distance and time.

Scattering is the change in the direction of photons caused by their collisions with other particles.

Fluorescence is the reemission of radiant energy (light) from a substance as a result of electromagnetic radiation or a stream of charged particles incident on the substance and being absorbed in it; when, upon removal of the incident energy, the emission of light takes longer than 10^{-8}s to decay, the phenomenon is known as *phosphorescence* rather than fluorescence, and the light-emitting substance is called a *phosphor.*

Opacity is the reciprocal of transmittance.

Turbidity is the cloudiness in a fluid due to the presence of suspended powdered or granular solids.

The *nephelos* of a substance is its ability to reflect light from within itself due to optical discontinuities in the substance, such as may be caused by a relatively weak suspension of particles in it.

1.2 CHEMICAL PROPERTIES AND COMPOSITION OF SUBSTANCES

Chemical properties include primarily the characteristics of solutions: their *concentration,* their oxidation-reduction potential (*ORP*), and their alkalinity or acidity (*pH*). Electrometric methods of analysis (see Chapter 3) are most commonly used for determinations of these properties. Most of the instruments described in the other chapters, however, are used for determinations of chemical composition.

Compositional analyses can be *qualitative* (What elements are contained in a sample?) or *quantitative* (How much of one or more given element is contained in a sample?). Either type can be a *single-component analysis* or a *multicomponent analysis.* An example of the purpose of a single-component qualitative analysis could be to determine if there is any cadmium in the sample; whereas if this sample were subjected to a quantitative analysis, the purpose would be to find how much cadmium is in this sample or the percentage of cadmium in the sample. Similarly, multicomponent analyses can be qualitative or quantitative, and they can be directed at finding two or more elements in a sample (or the quantity of those elements) or at finding all elements that compose a sample (or finding the quantities of all those elements).

When a sample of a process stream (or other fluid) is to be analyzed, one of two sampling methods can be employed: *in situ sampling* (the analyzer is located in, or immediately at, the process stream), or *extractive sampling* (a sample is extracted from the process stream and transported to an analyzer located elsewhere). Recent technological developments have led to another method: analysis by *remote sensing* (from ground-based equipment or from satellites).

The types of substances that can be found as a result of a compositional analysis can be elements in their *atomic* form (e.g., O, N) or *molecular* form (e.g., O_2, N_2); they can be more or less abundant *isotopes* of an element (with the same *atomic number* but a different *mass number*); they can be *inorganic compounds* or *organic compounds* (compounds based on carbon chains or rings, also containing hydrogen with or without oxygen, nitrogen, or other elements); and they can be solids, liquids, or gases, or *two-phase* liquids (e.g., oil + water), or mixtures, slurries, suspensions, or solutions. (Please refer to Appendix A for a listing of the elements and their characteristics.)

The development of many types of analysis instruments received a new impetus when water- and air-pollution controls went into effect. Some of the *acronyms* which became popular due to such extended use of such instruments are the following:

BOD: biochemical oxygen demand
COD: chemical oxygen demand
ORP: oxidation-reduction potential (redox potential or redox)
TC: total carbon
TOC: total organic carbon
NO_x: nitrogen oxides (NO, NO_2, NO_3)

1.3 UNITS, SYMBOLS, AND UNDERLYING CONCEPTS

1.3.1 Pressure

Pressure is expressed in units of force per unit of area, e.g., N/m^2; in the special case of pressure the N/m^2 is called the *pascal* (*Pa*), and the Pa is now the SI unit of pressure. A large variety of different units have been used to express pressure (see the subsequent table of conversion factors); relatively low pressures, for example, have been expressed in terms of liquid head (column height), such a millimeters of mercury (at 0°C)–a unit also called *torr* in vacuum work—or centimeters of water (at 4°C). A frequently used unit, the *normal atmosphere,* representing the ambient atmospheric pressure at earth sea level, was defined as the pressure indicated by a 760-mm-high column of mercury (at 0°C, at a density of 13.595 g/cm^3, and at an acceleration due to gravity of 980.665 cm/s^2). The most commonly used unit of pressure in the United States is the *pound force per square inch* (*psi*), which has been in use in more specific forms denoting what pressure the expressed pressure is referenced to: *psia* (for *absolute*), *psig* (for *gage*), and *psid* (for *differential*) pressure; such a differentiation is often very important, and no equivalent suffixes have been assigned to the Pa nor are any expected. It is, therefore, important to remember that Pa is a unit of absolute pressure unless otherwise stated (e.g., "tires are typically inflated to a *gage* pressure of 200 kPa," or "resulting in a *differential* pressure of 2.4 MPa"). Since the Pa is a very small unit of pressure (about 1.45×10^{-4} psi), its acceptance will meet considerable resistance, except for low-pressure measurements, especially in the United States. However, decimal multiples of the Pa offer more reasonable conversion magnitudes: 1 psi = 6.894 757 kPa or, where a very coarse approximation is adequate, 7 kPa; 1 kPa = 0.1 450 377 psi, or very close to 0.145 psi. Megapascals (MPa) can be used for higher and very high pressures; for example, 1 MPa ≈ 145 psi. Finally, in weather reports, atmospheric pressure is still very often expressed in millibars (mbar), where 1 mbar = 100 Pa = 1 hPa (*hectopascal*). Some countries have already adopted the hectopascal as a unit for atmospheric pressure and are using this unit daily in their official weather reports.

The following table gives conversion factors from various units of pressure to pascals:

To Convert from	to	Multiply by
atmosphere (normal)	Pa	$1.013\ 25 \times 10^5$
bar	Pa	10^5
dyne/cm^2	Pa	10^{-1}
foot of water (39.2°F)	Pa	2.989×10^3
inch of mercury (0°C)	Pa	3.3864×10^3
inch of water (4°C)	Pa	2.491×10^2
kilogram-force/cm^2	Pa	$9.806\ 65 \times 10^4$
kilogram-force/$meter^2$	Pa	9.806 65
millibar	Pa	100
millimeter Hg (0°C)	Pa	133.32
pound-force/$foot^2$	Pa	47.88
pound-force/$inch^2$	Pa	6.895×10^3
torr	Pa	133.32

Table B-1 (in Appendix B) gives quick conversions from psi to kPa and MPa.

1.3.2 Flow

Flow rate is expressed either in units of volume per unit time (volumetric flow rate) or in units of mass per unit time (mass flow rate). The SI units are the m^3/s for volumetric flow rate and the kg/s for mass flow rate. The mass flow rate of gases is often expressed as equivalent volume flow at standard conditions, i.e., 20°C and 1 standard atmosphere. When this is the case, the units are called *standard* m^3/s (or standard other units that can be converted to m^3/s).

Conversion factors

To Convert from	to	Multiply by
$foot^3$/minute	m^3/s	4.7195×10^{-4}
$inch^3$/minute	m^3/s	2.732×10^{-7}
$yard^3$/minute	m^3/s	1.2743×10^{-2}
gallon/minute	m^3/s	6.309×10^{-5}
liter/second	m^3/s	10^{-3}
pound-mass/second	kg/s	0.4536
pound-mass/minute	kg/s	7.56×10^{-3}

1.3.3 Humidity and Moisture

Relative humidity is expressed in *percent* (%RH).

Absolute humidity is usually expressed in units of mass per unit volume (kg/m^3, or, more commonly, g/m^3).

Moisture is expressed in percent by volume or percent by weight of either the total or the dry weight.

The *dew point* is expressed in units of temperature, typically, °C.

1.3.4 Density and Viscosity

Viscosity is expressed in units of force per unit area per unit time, e.g., $N \cdot s/m^2$.

To convert from *centipoise* to $N \cdot s/m^2$, multiply by 10^{-3}.

To convert from *slug/foot second* to $N \cdot s/m^2$, multiply by 47.88.

Kinematic viscosity is expressed in units of area per unit time, e.g., m^2/s.

To convert from *centistoke* to m^2/s, multiply by 10^{-6}.

To convert from *foot²/second* to m^2/s, multiply by 9.3×10^{-2}.

Density is expressed in units of mass per unit of volume, e.g., kg/m^3.

Specific gravity is typically expressed in the form t_f/t_s *specific gravity,* where t_f is the temperature of the fluid under consideration and t_s is the temperature of the fluid considered as standard (e.g., the 20/20°C specific gravity); it is a dimensionless number since it is a density ratio.

To convert from *gram/centimeter*^{3} to kg/m^3, multiply by 1000.

To convert from *pound mass/inch*^{3} to kg/m^3, multiply by 2.77×10^4.

To convert from *slug/foot*^{3} to kg/m^3, multiply by 515.4.

1.3.5 Liquid Level

Liquid level is usually expressed as a length dimension, the height of the liquid surface relative to a reference point.

Related measurements can be inferred from liquid level measurements by calculations of a type that can be handled easily by a microprocessor. Thus, when the geometry and dimensions of a tank are also known, the *volume* of the liquid can be determined. If, additionally, the density of the liquid is known, the *mass* of the liquid can be established.

1.3.6 Temperature and Heat

Temperature scales were originally established on an arbitrary basis; much later, international agreement was reached on an appropriately defiined temperature scale. The *Fahrenheit* scale is based on the mercury-in-glass thermometer, with the ice point defined as 32°F and the *steam point* (boiling point of water) defined as 212°F; the ice point–steam point difference is 180°F. The *Rankine* scale is an absolute temperature scale, with the same difference (180°R) between ice point and steam point as the Fahrenheit scale, but with the *absolute zero* of temperature defining 0°R; hence, the ice point is 491.7°R and the steam point is 671.7°R. Since both of these temperature scales have the same difference in degrees between ice point and steam point, 1°F = 1°R. The *Réaumur* scale was used in some European countries, but is now essentially obsolete; on this scale the ice point was 0° and the steam point was 80°. The *Celsius* scale is also based on the mercury-in-glass thermometer, but with the ice point defined at 0°C and the steam point at 100°C. This scale is widely used, and the °C is considered a unit of the metric system.

The unit of temperature in the International System of Units (SI) is the *kelvin* (K). It should be noted that the word "degree," or its symbol, is not used in conjunction with this unit. The kelvin is the unit of *thermodynamic* temperature and is defined as the fraction 1/273.16 of the thermodynamic temperature of the triple point of water. The triple point of water is 0.01 K above the ice point; hence, on the kelvin scale, the ice point is 273.15 K and the steam point is 373.15 K. The difference between these two temperatures is 100.00 K, and 1 K = 1°C, exactly. Absolute zero is 0 K.

The following relationships are used for temperature scale conversions:

°C = (°F − 32) × 5⁄9 and °F = 9⁄5 × °C + 32
K = °C + 273.15
°R = °F + 459.67 (which can usually be rounded off to 459.7)

Table B-2 (in Appendix B) is a convenient scale conversion table.

Heat flux is the time rate of transfer of heat energy. Energy is expressed in joules (*J*) and time in seconds (*s*); hence, heat flux is expressed in J/s or watts (*W*), since the watt, the unit of power, is equal to J/s. Other units (non-SI units) have been used to express heat power as well, and their conversion factors are 1 W = 1 J/s = 9.498×10^{-4} Btu/s = 10^7 ergs/s = 0.2389 g·cal/s. Conversion factors for the Btu (British thermal unit) are: 1 Btu = 8.139×10^{-5} W·s = 1055 J = 252 g·cal = 1.055×10^{10} ergs.

Because heat flux is measured as heat flow to a surface, it is also expressed in units of (radiant, conductive, or convective) *heat flux per unit area,* that is, in W/m². Other (non-SI) units are still in popular use in the United States, and the most commonly used unit has been the Btu/ft²·s (Btu per square foot per second). Conversion factors for this unit are: 1 Btu/ft²·s = 3600 Btu/ft²·h = 0.00695 Btu/in²·s = 0.271 cal/cm²·s = 975 cal/cm²·h = 1050 W/ft² = 7.3 W/in² = 1.136 W/cm² = 1.136×10^4 W/m².

Units for heat flux as well as the other heat-related quantities are tabulated as follows, together with their conversion factors:

Quantity	U.S. Customary Unit[a]	Multiply by	to get SI Unit
Heat flux	Btu/ft²·s	1.136×10^4	J/m²·s
Heat flux	cal/cm²·s	4.184×10^4	W/m²
Heat energy	Btu	1.055×10^3	J
Heat quantity	cal	4.1868	J
Thermal conductivity	Btu·in/s·ft²·°F	5.192×10^2	W/m·K
Heat power	Btu/h	0.293	W
Specific heat	Btu/lb_m·°F	4.1868×10^3	J/kg·K
Heat-transfer coefficient	Btu/s·ft²·°F	2.044×10^4	W/m²·K
Thermal resistance	°F·h·ft²/Btu	0.1761	K·m²/W
Thermal diffusivity	ft²/h	2.58×10^{-5}	m²/s

[a]Conversion factors for *British thermal units* (*Btu*) and *calories* (*cal*) are based on the International Tables; conversion factors for *thermochemical Btu* and *cal* differ from those shown by about 0.07%.

1.3.7 Optical Properties

Luminous intensity (*of a point source*). The unit of luminous intensity is the *candela* (*cd*), one of the base units of the SI. It is defined as the luminous intensity, in the perpendicular direction, of a surface of 1/600 000 m^2 (1/60 cm^2) of a black body at the temperature of solidifying (freezing) platinum under a pressure of 101 325 Pa. A proposed redefinition of the candela is the luminous intensity in a given direction of a source emitting a monochromatic radiation of frequency 540×10^{12} Hz and whose radiant intensity in that direction is 1/683 W/sr.

An older unit, still popular, is the *candlepower* (*cp*), defined as the luminous flux of one *candle* (see luminous flux, next) when viewed in the horizontal plane. There is a one-to-one relationship between the old and new units (1 cp = 1 cd).

Luminous Flux. The unit of luminous flux is the *lumen* (*lm*), defined as the flux emitted within a solid angle of one steradian by a point source having a uniform intensity of 1 candela (lm = cd·sr). An older unit was the *candle,* the luminous flux from a spermaceti candle burning at the rate of 120 grains per hour.

Illumination (*of an area*). The unit of illuminance is the *lux* (*lx*), the illumination of one lumen per square meter (1 lx = 1 lm/m^2). An older unit was the *footcandle* (*fc*), the illumination at a spherical distance of 1 foot from a 1-candle source (1 fc = 10.76 lx).

Brightness (*Luminance*). The unit of luminance is the *candela per square meter* (*cd*/*m^2*) (of light-emitting area). An older unit is the *footlambert* (*fL*), the luminance equal to $1/(4\pi)$ candle per square foot (1 fL = 3.426 cd/m^2). Among other units that have been used are the *lambert* (*L*), the luminance equal to $1/\pi$ candle per square centimeter (1 L = 3183 cd/m^2) and the *stilb* (*sb*), the brightness equal to 1 candle per cm^2 (1 sb = 10 000 cd/m^2).

Wavelength. Wavelength in the UV and visible portions of the spectrum is expressed in *nanometers* (*nm*), whereas it is usually expressed in *micrometers* (*μm*) in the IR region. The micrometer has previously often been referred to as a micron (μ). Another older unit, whose use is now discouraged, was the ångstrom (Å, or just A), which is equal to 0.1 nm (10 Å = 1 nm).

Other Quantities. Luminosity is expressed in *lumens per watt* (*lm*/*W*).

Black-body temperature is expressed in *kelvins* (*K*).

Photon energy is usually expressed in *electron volts* (*eV*) or *ergs;* the SI unit is the *joule* (*J*) (1 eV = 1.602×10^{-19} J; 1 erg = 10^{-7} J).

Wave number is expressed in cm^{-1}.

Nonvisual magnitudes. Radiant flux is expressed in *watts,* radiant intensity in *watts per steradian,* radiant flux per unit area in *watts per square meter,* and radiance in *watts per square meter per steradian* ($W\ m^{-2}\ sr^{-1}$).

1.3.8 Chemical Properties and Composition

The SI unit of substance is the *mole* (*mol*), the amount of substance of a system which contains as many elementary entities as there are carbon atoms in 0.012 kg of carbon-12 (C^{12}); the elementary entities must be specified and may be atoms, molecules, ions, electrons, other particles, or specified groups of such particles.

Atomic weight (*at wt.*) is the relative mass of an atom based on a scale on which a specific carbon atom, C^{12}, is assigned the mass value 12.

Molecular weight is the sum of the atomic weights of all the atoms in a molecule.

Mole fraction is the ratio of the moles of a given substance, in a mixture or solution, to the total number of moles of all components in that mixture or solution.

Gram-equivalent weight (gram-molecular weight) is the equivalent weight of an element or compound expressed in grams on a scale in which carbon-12 has an equivalent weight of 3 grams in those compounds in which its formal valence is 4.

An *atomic mass unit* (*amu*) is an arbitrarily defined unit in terms of which the masses of individual atoms are expressed; the standard is the unit of mass equal to $1/12$ the mass of the carbon-12 atom (the carbon atom having as nucleus the isotope with mass number 12).

The quantity of a single component in a mixture or solution is commonly expressed in *percent* (*%*), or in *parts per million* (*ppm*) or *parts per billion* (*ppb*), where a billion is 10^9; in order to be more consistent with the SI, the latter two units could be expressed as . . . $/10^6$ and . . . $/10^9$, respectively.

The ratio of unit mass to unit volume is often used in compositional analyses, typically expressed in *micrograms per milliter* ($\mu g/ml$), or in $\mu g/m^3$.

The frequency of spectral lines or bands, in optical and quasi-optical spectrometry, is usually expressed as *wavenumber*, in cm^{-1}; the wavenumber is the reciprocal of the wavelength, with the wavelength expressed in centimeters.

The *atomic number* of an element is the number of protons in its nucleus.

The *mass number* of an element is the sum of the numbers of protons and neutrons in its nucleus; it is commonly written as a superscript before (or sometimes, after) the symbol of the atom, e.g., ^6Li (preferred), or Li^6.

2

INTRODUCTION TO INSTRUMENTS

2.1 INSTRUMENT SYSTEMS

2.1.1 Instrument, Sample, and Sensing Fundamentals

Electronic instruments used for the determination of primarily chemical properties and composition are utilized for information (analysis) as well as for control purposes. When used to obtain information, they are frequently used individually, but are also used, together with other instruments, in an *information system* (*data system*). How the information is used is basically up to a human observer.

For control, instruments are most commonly used, together with other devices and various manual controls, in a *control system.* An increasing number of control systems are of the *closed-loop* type, where a human operator adjusts settings on electronic equipment (selects *set-points*), and the system then responds to data from the instrument(s) to adjust a control device, such as a valve, to maintain the analyzed parameter at the set-point or within a selected range of set-points.

The information is obtained by the instrument from a *sample.* As indicated in Figure 2-1, the sample can be liquid, solid, or gaseous, or it can be a mixture of these states. For some of the illustrations in this book, a generalized representation of a sample was selected. The nature of the sample can, of course, span a very wide range. To cite a few examples, the sample can be a process fluid (liquid or gas), the ambient atmosphere (e.g., one containing, or suspected of containing, toxic, combustible, or polluting gases), waste water, or a slurry; it can be plastic, paper, textiles, mineral, or metal; it can also be a planet or its atmosphere, or even a remote star or galaxy.

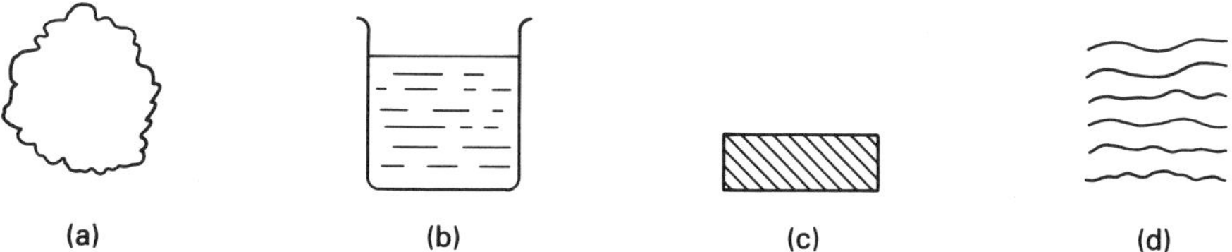

Figure 2-1. The sample: (a) generalized; (b) liquid; (c) solid; (d) gaseous.

A majority of instruments are now used for various determinations of characteristics of fluids (a fluid can be liquid or gaseous). Figure 2-2 shows typical methods of fluid sensing. A sample can be extracted from a liquid and then poured into a beaker for analysis (usually in a laboratory); similarly, a gaseous sample can be extracted and then injected into a port of an instrument. Trends to install instruments *in situ* are increasing, so that their sensing portion is exposed directly to a process fluid or a process by-product such as stack emissions. A sensor or detector is then mounted directly into a port on a pipe or tank (Figure 2-2b), or, if such direct exposure is deemed undesirable, into a by-pass line (Figure 2-2c). In some cases the sample needs to be conditioned in some manner before it can be analyzed. A sample extraction line can

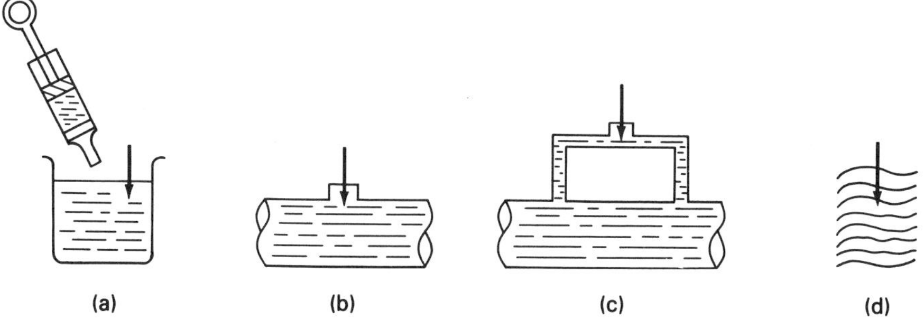

Figure 2-2. Methods of fluid sensing: (a) extractive sample; (b) in situ; (c) in situ with bypass; (d) noncontained.

then be connected to the process pipe or container; this line leads to conditioning equipment from which it can, in most cases, be directly connected to the analyzer. Finally, for many analyses of ambient atmospheres, the sample is noncontained and the sensor or detector is simply exposed to the atmosphere.

Sensing techniques used for analyses can be categorized as passive or active. Both of these categories can employ direct sensing, noncontact sensing, or remote sensing (see Figure 2-3). In *direct sensing* the sensing portion of the instrument is immersed into the sample or pressed against it. In *noncontact sensing* the sensing portion is positioned relatively close to the sample, but not in physical contact with it. In *remote sensing* the sensor can be several meters, or tens, hundreds, thousands, millions, or even billions of kilometers away from the sample. Many types of instruments employ passive sensing. *Active sensing* involves adding a source of a beam (typically electromagnetic energy) that interacts with the sample. This source is usually built into the instrument system.

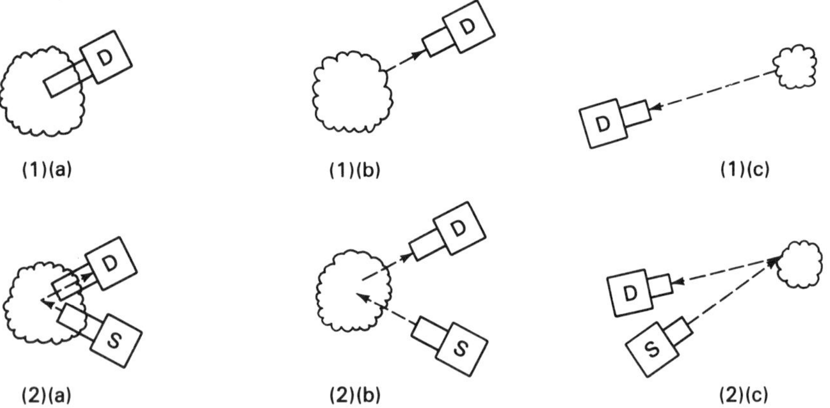

Figure 2-3. General types of sensing: (1) passive; (2) active; (a) direct; (b) noncontact; (c) remote; (D = detector, S = source).

2.1.2 Instrument System Elements

A basic instrument system is shown in Figure 2-4a. It consists of

1. The *sensor* or detector, which responds to a given characteristic of the sample with a corresponding electrical output.
2. The *signal conditioner,* which converts this output into the type of electrical signal that the display device will accept.
3. The *display device* (or read-out device), which displays the required information about the sample characteristics.

4. The *power supply,* which converts the power from the power source (battery or power line) into voltages required by the signal conditioner, by most types of sensors or detectors, and by many kinds of display devices.

A fifth function, not considered an instrument element, is *data interpretation,* performed usually by a human operator, sometimes by a closed-loop control system. Data interpretation means examining the displayed information in the context of the purpose of the analysis.

Sensors and detectors are specific to a given type of instrument and are covered in subsequent chapters of this book. Signal conditioners can vary in complexity from relatively simple passive networks to elaborate active circuitry. Amplifiers are very often used for signal conditioning. They can be linear or logarithmic, amplify ac or dc signals, be connected as bridge, differential, balanced, or single-ended amplifiers, and provide various levels of gain. Filters, sometimes high-pass or band-pass but more often low-pass, are often included in signal conditioners. Analog-to-digital converters (ADCs) are increasingly incorporated in signal conditioners so that a digital read-out (display) can be used, or to facilitate a computer interface. Additional functions are often included in more elaborate instrument systems.

Display devices also vary in complexity. Simple d'Arsonval (moving-needle) meters (which need not be connected to a power supply) are being phased-out in favor of digital read-outs. Other display devices include circular-chart and strip-chart recorders, (digital) character printers, and cathode-ray tubes (CRTs). Visible and/or audible alarm indicators are sometimes added to the display device.

A basic analysis instrument system (see Figure 2-4b) differs from a basic instrument system in that an analysis section is added. This section often has the form of additional circuitry; however, it can also be a scanning device such as a monochromator that is ahead of the detector, or it can be circuitry, added between power supply and sensor, that, for example, varies sensor electrode potentials. If the instrument system employs the active sensing technique, a source (usually of a beam) is also added. In some instruments, the source is designed to be controlled in such a manner that it aids, or even plays a primary role, in the analysis function.

With the increase in availability and reduction in cost of many types of microprocessors, such devices are now often incorporated in analysis instrument systems. They can be used to process data in various ways, as well as to control a variety of functions within the instrument; this control is only sometimes (and then only partly) autonomous; the controls are primarily actuated in response to external commands issued by a human operator. The microprocessor also often acts as part of the interface with a data bus (data highway) used in multi-instrument data systems and control systems.

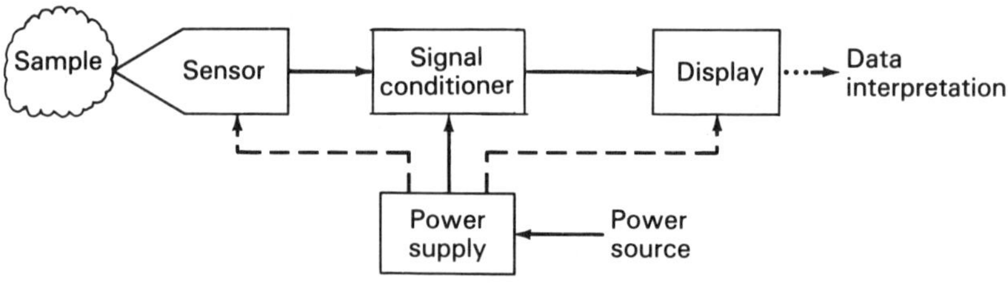

(a) Basic instrument system

Sample
*)
Sensor or detector
Signal conditioner
Analysis section
Display
Data interpretation
*)
*)
Source
Power supply
Power source

(b) Basic analysis instrument system
(* —alternative or additional analysis device)

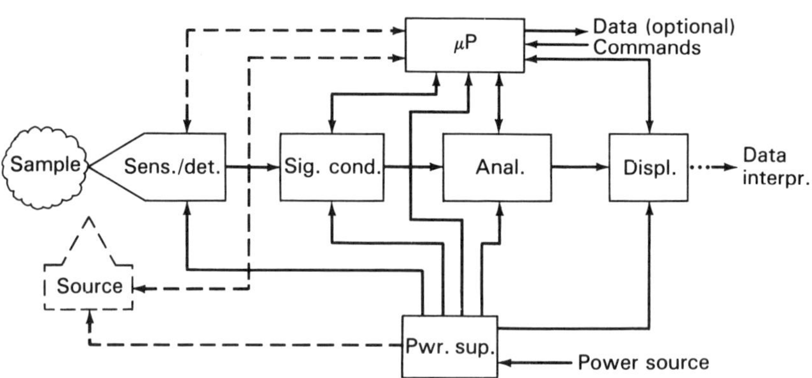

(c) Analysis instrument system with microprocessor

Figure 2-4. Evolution of electronic analysis instruments.

Since microprocessors are digital in nature, the signal conditioner of instruments incorporating them always includes an ADC. The ADC converts the amplitudes of conditioned signals into digital form, usually data words of specified length and typically encoded in natural binary, sometimes in binary-coded decimal, form. The number of (binary coded) bits in a data word depend on the resolution of signal amplitude needed in displayed data (see

Table 2-1). The output of the ADC is usually a serial digital data stream. There are many ways in which a microprocessor can be used in conjunction with an ADC to facilitate the manipulation of data for the purpose of simplifying or accelerating data interpretation. For example, it can provide for zero suppression where only signal amplitudes above a specified level are to be displayed, for (programmable) alarm (out-of-limit) displays, for amplifier gain changes and gain status indications as part of a data word, for suppression of data that do not change (or even whose rate of change does not vary) by a specified amount, and for temporary storage and display of signal-fluctuation peaks or valleys (*peak-hold, valley-hold*).

TABLE 2-1. Resolution of Digitized Analog Measurement as Function of Length of Digital Word

Number of Discrete Increments	Word Length (bits)
1	1
3	2
7	3
15	4
31	5
63	6
127	7
255	8
511	9
1023	10
2047	11
4095	12
8191	13
16 383	14
32 767	15
65 535	16
131 071	17
263 143	18
524 287	19
1 048 575	20
2 097 151	21
4 194 303	22
8 388 607	23
16 777 215	24
33 554 431	25

2.2 INSTRUMENT DATA SYSTEMS

Data systems employing more than one instrument can be categorized as follows: *information systems* are intended solely for the display of information about samples; *control systems* are aimed primarily at control of various types of processes on the basis of data received from instruments, with control functions operator-initiated, operator-assisted, or autonomous, or a combination of these.

2.2.1 Information Systems

In multiple-instrument information systems the display function is removed from each instrument system and arranged in such a manner that data from all instruments can be displayed on one or more central display devices. The signal conditioning function is almost invariably retained within each instrument.

Simple information systems, such as those illustrated in Figure 2-5, are used mainly with relatively simple instruments (e.g., pH and conductivity sensors) which represent the characteristic to be analyzed by a single analog output. If the calibration of each of the sensors differs significantly from a nominal calibration, the signal conditioners can be adjusted so as to standardize all the outputs. In the simplest of these systems (Figure 2-5a), a manually-operated selector switch, with switch positions marked on the panel, is used to select the instrument to be read out. This manual operation can be automated by using a stepper switch which sequentially connects conditioned sensor outputs to the display device, while simultaneously displaying the switch position next to the read-out (Figure 2-5b). An ancillary printer can be used to obtain a permanent record, provided the read-out device is equipped for connecting it. A display device such as a multichannel strip-chart recorder (oscillograph) can display the outputs of several instruments simultaneously; this is particularly useful if variations, with time, of several characteristics need to be compared with each other (Figure 2-5c).

There are many types of information systems whose complexity is somewhere between that of the simple systems just described and the distributed data system illustrated in Figure 2-6. This system serves the processing and display of information from a number of instruments, each producing data in digital form. The processing, formatting, etc. of instrument data is performed by a computer (a central processing unit, CPU, and memory). The computer is usually controlled from a dedicated control operator station, which also provides for programming the computer (using software), as well as diagnostics and maintenance. Appropriate types of cabling constitute the network, which permits bi-directional communications. The network should be designed to meet an accepted industry standard. The instruments interface with this network through an interface unit (NIU); in some cases this NIU is built into the instrument. Each of the display and ancillary devices is interfaced with the network by an NIU (sometimes referred to as access control unit or network access controller). Among the ancillary devices can be a complete

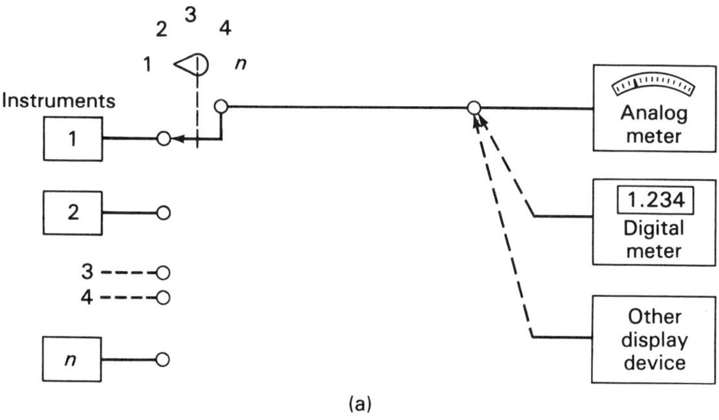

(a)

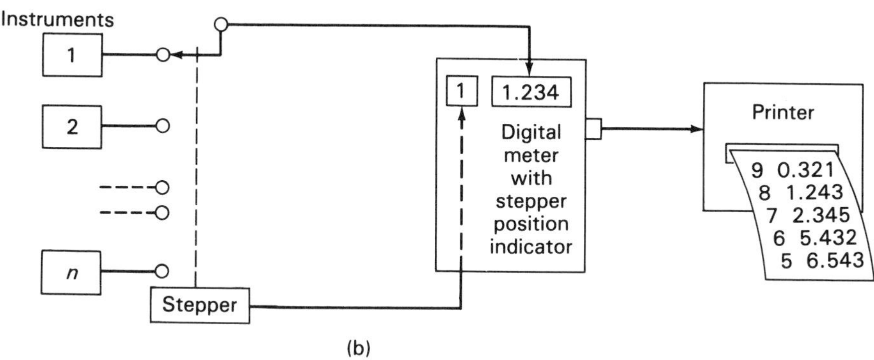

(b)

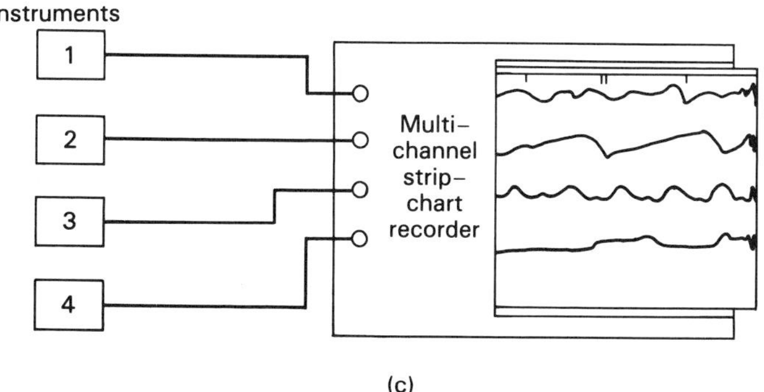

(c)

Figure 2-5. Simple multiple-data information systems: (a) manually selected measurements; (b) automatically selected measurements; (c) simultaneously displayed measurements.

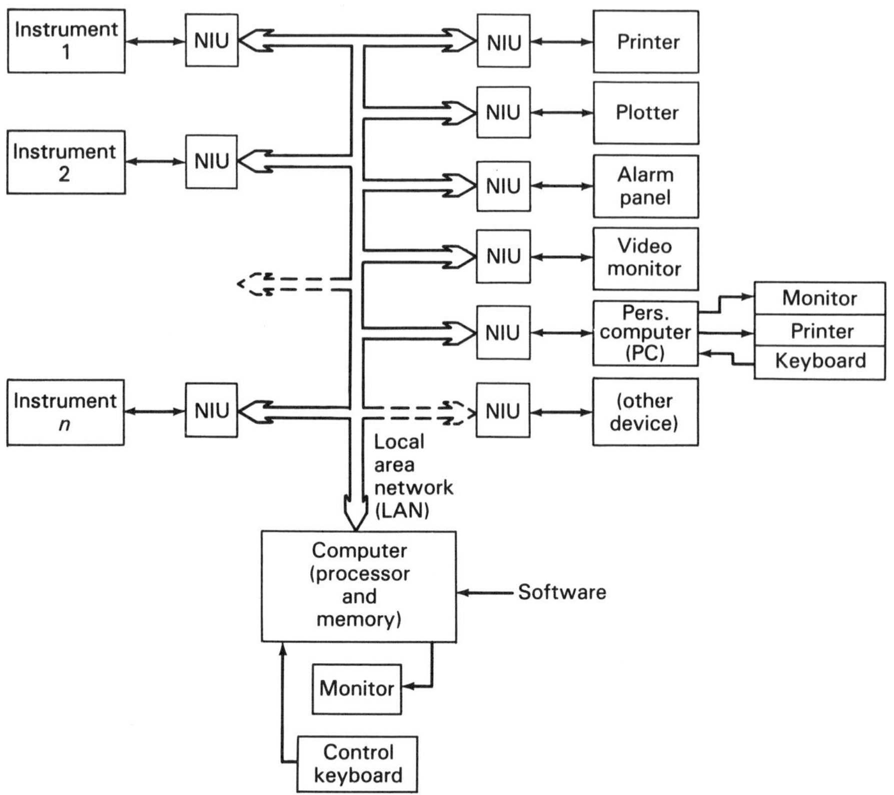

Figure 2-6. Distributed instrument information system (Note: NIU = Network interface unit).

personal computer, with keyboard, monitor and printer. Such a computer may also use additional software for special processing of data. As indicated in the illustration, other devices can also be interfaced with the network; examples of such devices are a modem (to interface with a telephone line) and a data storage device (e.g., tape recorder). More than one of each display or ancillary device can be connected to the network.

Telemetry systems are used when data are acquired at a remote site but must be displayed at a local site. The remote site is, generally, a location to which human access is difficult or impossible; it can be, for example, a section of pipeline in the desert or wilderness, an ocean buoy, a robotic vehicle, a satellite, or a planetary spacecraft. The local site would typically be an easily accessible major facility.

All telemetry systems use a modulated high-frequency carrier to transmit the information from one point to the other. Most such systems are *radio telemetry systems* which use transmitting and receiving antennas and a carrier at radio frequencies (long-wave to microwave). Some telemetry systems are *carrier-current systems* in which the modulated carrier is coupled directly to an existing line (e.g., a power line) and then decoupled from this line at the receiving end.

A generalized basic radio telemetry system is illustrated in Figure 2-7. Instrument data are fed to a *multiplexer* (commutator) which combines them into a single *composite* signal. This signal is applied to the high-frequency *transmitter* where it modulates the output of an oscillator. The modulated *carrier* is then amplified and fed to the transmitting antenna. This antenna, which can be highly directional, usually is pointed to radiate the modulated carrier toward the receiving antenna. The received signal is first amplified and then applied to a *demodulator,* which strips the carrier from the composite signal. A *decommutator,* which must be synchronized with the commutator at the transmitting end in some manner, is then used to extract individual instrument data from the composite signal. In some cases the instrument data can then be displayed directly; in other cases a *data processor* is used to achieve the desired type of display.

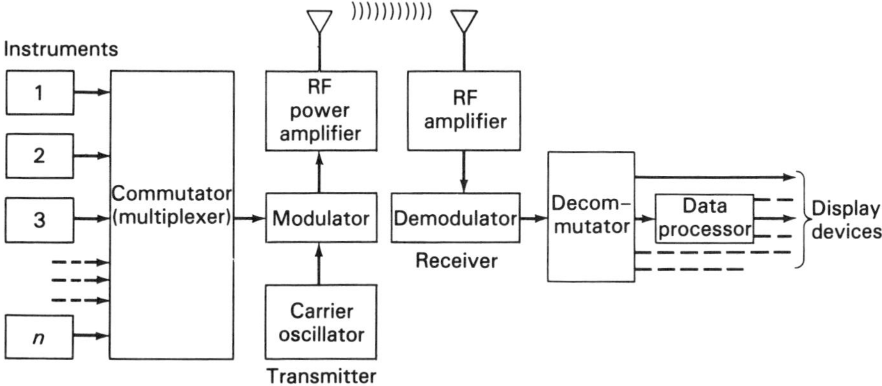

Figure 2-7. Basic radio telemetry system.

In earlier telemetry systems the instrument data applied to the multiplexer were in analog form, but standardized (by means of signal conditioning, if necessary) so that, for example, 0–5 V dc always represents the full-scale output of an instrument. The earliest multiplexer was an electromechanical multicontact switch (commutator) that rotated continuously at a constant rate. Modern telemetry systems are usually digital in nature. The analog-to-digital converter (ADC) can be a part of the signal conditioner of each instrument, or, as long as all instrument outputs are standardized (as explained above), a single ADC can be used to digitize the composite signal. Since digital data are on/off in nature ("ones and zeros"), the type of carrier modulation employed is now typically either frequency-shift keying (FSK), in which the frequency of the carrier is changed by a discrete increment or decrement whenever a "1" is encountered, or phase-shift keying (PSK), in which the phase of the carrier is changed similarly. In either scheme the amplitude of the carrier remains constant.

The type of multiplexing of instrument data is important to the manner in which data are displayed and can be interpreted. *Frequency-division multiplexing* involves the simultaneous modulation of a number of subcarriers (each of a different frequency range, but each much lower in frequency than the carrier); it allows the continuous display of a limited number of instrument outputs; it is rarely used any longer. The method most commonly used now is *time-division multiplexing* (*TDM*) in which each instrument output is sampled sequentially for a brief period of time. This method does not permit a continuous display of data from each instrument; however, the characteristic being analyzed can be adequately reconstituted in the display as long as the sampling rate is fast enough.

There are several levels of complexity between the basic telemetry system shown in Figure 2-7 and the advanced system illustrated in the simplified block diagram, Figure 2-8. This information system not only sends data from the remote to the local site; it also allows the instruments to be controlled by commands sent from the local site. Such commands can range from a simple "turn instrument on/off" to effecting state changes within an instrument, such as changing amplifier gain, electrode potentials, or optical filters. Each instrument in this advanced system is equipped with a microprocessor that performs a variety of functions; for example, it acts as interface with the data bus, it responds to command data by effecting state changes, it encodes analog data into digital data of a specified word length, it buffers instrument data in its memory, and it can add *headers* to the data, to identify the instrument and to indicate at exactly what time the data were acquired. If *packet telemetry* is used, each instrument produces a *source packet* of a specified length (expressed in bits). The header of each source packet includes instrument ID and clock time. This greatly simplifies data processing at the local site: the computer reads the instrument ID and then channels the data packet to an instrument-dedicated buffer, processor, and display.

The data handling and control (DHC) equipment performs the functions indicated in the block diagram. It also interfaces with the data bus through a microprocessor. Data from each instrument are acquired, typically by sending a "read out your buffer" command to each instrument. These read-out commands can be sent in a fixed sequence (as in time-division multiplexing), while

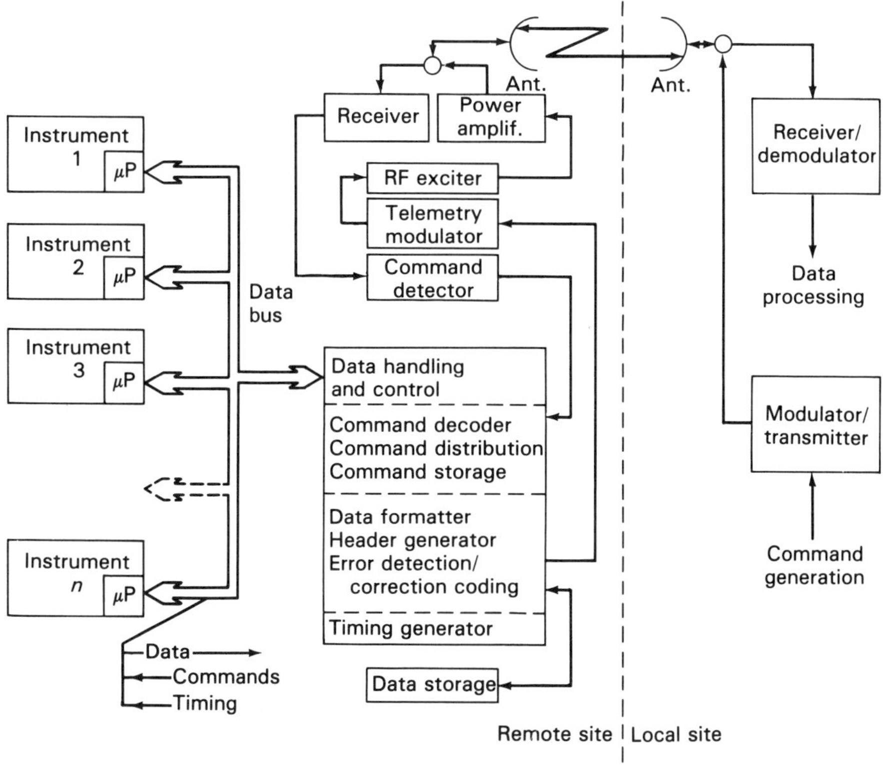

Figure 2-8. Advanced instrument telemetry system.

still allowing for the fact that some instruments produce more data in a given time interval than others. In situations where some instruments produce more than the average quantity of data in certain periods of time while some other instruments produce few or no data, this "fixed polling" can be replaced by "adaptive polling", where a given instrument sends the information "my buffer is full, read me out" to the DHC. The DHC then sends the read-out command to that instrument, but within the constraints of a programmed protocol. The DHC then formats instrument data into a *frame* (in packet telemetry it formats source packets into a *transport frame*); each frame starts with a frame header, and each frame header starts with a group of bits used for frame synchronization, so that the local-site data processor can tell when a new frame begins. Most such telemetry systems include data storage (typically a tape recorder), so that data are not lost when the radio link between the two sites encounters a problem; the tape recorder can later be commanded to play back the stored data. Finally, the CDH can apply error detection and correction coding, of a specified scheme, to each data frame.

2.2.2 Control Systems

Control systems can be considered as information systems with provisions added to respond to the information by effecting control functions. For example, a simple pH information system is used to display the pH of a liquid. The corresponding control system adds control equipment that maintains the pH at a specified value, typically by actuating control valves that reduce or increase the addition of acidic or alkaline liquid to the process liquid. The control system almost invariably requires a human operator to be in the loop. In an *open-loop* control system, the operator looks at the display and then either turns valves manually or induces the electrical or pneumatic action of the control valves from a control panel. In a *closed-loop* control system, the output of the pH sensor is used to effect the proper settings of the control valve(s) automatically. But even when closed-loop control is employed, a human operator should be available to verify that the automatic control function was properly implemented. Typically, though, automatic control systems are multifunction and the operator monitors many other control functions as well.

The most commonly used automatic control systems are closed-loop systems employing feedback. A feedback loop includes a forward signal path, a feedback signal path, and a summing point, which together form a closed circuit. A typical basic closed-loop control system is illustrated in Figure 2-9. It operates in the following manner (equivalent terms commonly used in process control are shown in parentheses).

A specific quantity within a *controlled system* (*process*) is to be maintained at a specified value. This *controlled quantity* (*controlled variable*) is

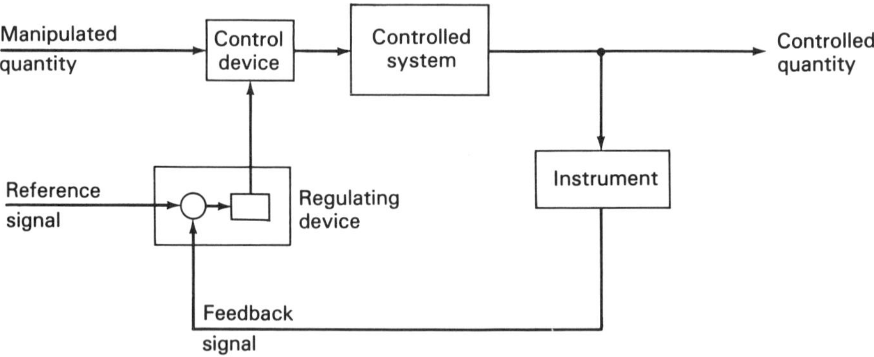

Figure 2-9. Basic closed-loop control system.

sensed by an instrument. The output of the instrument, which usually incorporates signal conditioning, is fed to a *comparing element* in a *regulating device* (*controller*) in which the instrument output signal (*feedback signal*) is compared with a *reference signal* (*set-point signal*). If the two signals are of the same magnitude, or within a relatively narrow tolerance from each other (*dead band*), no further action occurs. If the two signals differ from each other by an amount larger than that tolerance, a regulating signal (sometimes called *error signal*) is sent to a *control device* (*final control element*). This signal causes the control device to change a quantity or condition (*manipulated quantity* or *manipulated variable*) in the controlled system. The control action remains in effect until the controlled quantity or condition is at its proper level, as indicated by the feedback signal equaling the set-point signal.

There are many types of control action. The simplest is *on-off control,* as exemplified by the thermostatic furnace control used in most homes. When the room temperature drops below the set point, the furnace is turned on. When the temperature rises and reaches the set point again, the furnace is turned off. This control action usually results in a noticeable temperature change, particularly if the dead band of the controller is too wide. When the temperature of a heating device must be kept at a more constant level, as in many industrial applications, *proportional control* can be applied. This control action provides a continuous linear relation between the output and the input of the controller. A small deviation in the temperature from the desired value, as sensed by a temperature sensor, causes a small regulating action to restore the temperature to that level. A large deviation causes a large regulating action. In this manner the temperature is controlled more closely than by on-off control.

Among additional types of control action are *derivative control,* in which the controller output is proportional to the rate of change of the output, and *integral control,* in which the rate of change of the output is proportional to the input. The set point may be set manually, automatically, or in accordance with a program. Programmable controllers, called *programmable logic controllers* (*PLC*), are increasingly used in modern control systems.

Distributed control systems (*DCS*) are now used in many large installations. They are comparable to the system shown in Figure 2-6, with, of course, all the required control functions interfaced with the network.

2.3 INSTRUMENT CHARACTERISTICS

In general, instrument characteristics can be categorized into design, performance, and reliability characteristics. All of these should be considered when specifying an instrument or selecting one from manufacturers' catalogs and bulletins; however, not all of the characteristics explained below may be important or applicable to a given type of instrument. On the other hand, a few specialized characteristics are important only to one category of instruments; they are dealt with in the appropriate chapter of this book.

Some of the characteristics apply primarily to instruments that can also be called *sensors.* Significant examples of these are most of the electrometric (sometimes called electrochemical) sensors: conductivity, pH, ORP, and specific ion sensors and some types of gas analyzers. The term *transmitter* is used instead of *sensor* in most process control applications; this alternative term usually also implies that the transmitter has a current output, typically 4–20 mA full scale.)

2.3.1 Design Characteristics

The design characteristics of an instrument describe or specify how the instrument is designed and constructed (or how it should be, in a user-written specification) in regard to its measuring range and its electrical and mechanical characteristics.

The *range* of an instrument is given by the lower and upper limits of measurand values it is intended to respond to within specified performance tolerances (a *measurand* is the characteristic or quantity to be measured). A range can be *unidirectional* (e.g., 0 to 14 pH), or *bidirectional* (e.g., − 1000 to + 1000 mV ORP; it can also be *expanded* (e.g., 4 to 10 pH) or *zero-suppressed* (e.g., 5 to 14 pH). The algebraic difference between the two range limits is the *span* of the instrument.

The *overrange* (overload) is the maximum magnitude of measurand that can be applied to the instrument without causing a change beyond specified tolerances. The *recovery time* is the amount of time allowed to elapse after removal of an overrange condition before the instrument again performs within specified tolerances.

Electrical design characteristics apply primarily to the output of the instrument, its *excitation* (power supply) and the electrical interfaces between these, as well as those between the instrument and electronic equipment to which the output is fed.

Output is the electrical quantity produced by the instrument as a function of the measurand (or as a result of an analysis). *Analog output* is a continuous function, usually a voltage or (if the instrument is a "transmitter") a current. In some cases, the analog output can be in terms of frequency (*frequency output*), where the number of cycles or pulses per second are a function of the measurand; it can also be (typically bidirectional) deviation from a specified center frequency (*frequency-modulated output*). *Digital output* represents the measurand in the form of discrete quantities coded in some system of notation, typically binary code (natural binary code, NBC), sometimes binary decimal code (BDC). A switch-type instrument can be said to have a *discrete-increment output,* in the form of one "off" state, and either one "on" state or several "on" states, each for a different level of the measurand.

End points are the output values at the lower and upper limits of the range or span of an instrument; for example, if the specified output is 4 to 20 mA, then 4 mA is the lower end point and 20 mA is the upper end point; or, if the specified output is 100 to 1000 mV, then 100 mV is the lower end point and 1000 mV is the upper end point. Tolerances can be specified for end points; for example, if the nominal output of an instrument is 0 to 5 V dc, the end point tolerances could be specified as 0.00 ± 0.02 and 5.00 ± 0.03 V dc. The *terminal end points* for this example, however, would simply be 0.00 and 5.00 V dc. Terminal end points (which are zero-based) are a form of *theoretical end points,* which are not necessarily zero-based (e.g., 4.00 and 20.00 mV dc).

Power supplies (excitation supplies) are needed by all electronic instruments. In most cases, either a battery (long-life, or rechargeable for portable instruments, or the battery of a vehicle on which the instrument is mounted) or the existing power line is used to supply this power. The power supply voltage (and frequency for ac power) as well as the maximum power consumed by the instrument should always be specified.

The impedances across the terminals of instrument, power supply, and electronics (to which the output signal is fed) become important. The impedance of the power supply, including cabling to the instrument, is the *source impedance.* The impedance across the power-input terminals of the instrument is the *input impedance.* The impedance measured across the signal-output terminals of the instrument is the *output impedance.* The impedance presented to these output terminals by external circuitry to which they are connected (including interface cabling) is the *load impedance.* Significant errors can result from having poorly matched impedances, e.g., when the load impedance is much lower than the output impedance. For current-output instruments ("transmitters") the overall *loop impedance* (especially the maximum loop impedance), at a specified voltage, is critical. If the output is digital, the *source and sink* characteristics of the instrument's output circuitry and of the external circuitry to which the output is fed are important.

Shielding and grounding of interfacing cabling deserve special consideration. It is usually good practice to isolate power ground and signal ground from each other, and to isolate both from case ground (the metallic instrument housing). The instrument housing should not be used as electrical ground. Depending on the grounding philosophy of the system, the housing can or can not be electrically connected to the shield of shielded cabling; some instrument contain a guard shield around a group of noise-sensitive components, and this (otherwise ungrounded) separate shield should be connected to the shield of a cable. The resistance between signal (and power) grounds and the housing is the *insulation resistance* of the instrument, usually expressed in megohms; it can also be expressed as the *breakdown voltage rating*, the voltage that, if applied across these grounds, causes the insulation to break down (fail). Proper isolation is important to common-mode rejection.

Other electrical design characteristics include those associated with signal conditioning: noise, ripple (undesirable line-frequency ac components superimposed on a dc signal), harmonic content (undesirable ac, representing harmonics of the line frequency, superimposed on a dc output signal), and amplifier gain instability. Ripple and harmonic content can also be caused by an internal oscillator instead of the power line.

The identification of external electrical connections are covered below. If the instrument consists of two or more elements, the interconnecting cables should be defined.

Mechanical Design Characteristics of an instrument always include the mass (weight); if an instrument is composed of separate sensor and electronics units, as well as mandatory ancillary equipment, the mass of each should be shown. Equally important is the *configuration* of the instrument, best shown on a drawing that includes all pertinent dimensions and the location, type, size, and orientation of all external mechanical, electrical, and fluid connections, including mounting holes or other mounting provisions. Again, if the instrument consists of two or more elements, these requirements apply to each element. If a sensor (probe) incorporates any special adjustment provisions (e.g., zero and gain or span adjustments), these should also be identified on the drawing. In some cases the dimensions of an enclosure can be defined by an accepted government or industry standard; this applies as well to electrical and fluid connectors. The materials and finish of housings should be identified. For sealed sensors, the type (and, if applicable, the pressure rating) of the seal should be stated. The materials that come in contact with a measured fluid should also be identified (see Appendix C).

Some applications require that the instrument meet certain industrial or governmental standards, for example if they must be capable of operating in a hazardous environment, must be explosion proof, must be intrinsically safe, or must be waterproof. The codes or standards that the instrument can meet are then stated.

Identification (nameplate information) is another important mechanical characteristic. As a guideline, nameplate information should be sufficiently complete for a user to be able to install and connect an instrument without looking up manufacturers' drawings or bulletins. Thus, the nameplate should show proper nomenclature for the instrument as well as its most pertinent characteristics such as range, supply power, and output, and, of course, the manufacturer's name and location, part number, and serial number. When an instrument is a self-contained sensor or probe, the electrical connections (connector pins or screw-terminals) should be identified as to their function (e.g., pin A—117 V ac, 50/60 Hz in; pin B—power return; pin C—signal out; pin D—signal return). Additional information may be required on the nameplate by applicable standards or codes, or by special user requirements. When something can be mishandled or misconnected, warning labels should be affixed at an appropriate location.

Display characteristics as well as the characteristics of ancillary equipment such as sample acquisition and conditioning are instrument-system dependent and are not itemized here.

2.3.2 Performance Characteristics

The performance characteristics of instruments can be categorized as follows.

- **a.** *Static characteristics* describe performance at room conditions (25 ± 10°C, 90% or lower relative humidity, and an ambient atmospheric pressure of 88 to 108 kPa [880 to 1080 mbar]), with very slow changes in the value of the measurand, and in the absence of shock and vibration.
- **b.** *Dynamic characteristics* relate to the response of the instrument to fluctuations or step changes in the value of the measurand.
- **c.** *Environmental characteristics* describe the performance (or changes from the performance at room conditions) of the instrument to specified external conditions such as temperature (ambient or that of the measured fluid), shock, vibration, and humidity.

For electronic analysis instruments, the major static characteristics comprise accuracy, repeatability, hysteresis, linearity, sensitivity and its shift, zero-measurand output and its shift (zero shift), resolution, and chemical selectivity. The primary dynamic characteristics are transient response (response time, rise time, time constant) and, in some cases, frequency response. Environmental characteristics are typically limited to temperature error and resistance to shock and vibration (resistance to hazardous and corrosive atmospheres is considered a mechanical design characteristic).

Many of these characteristics are expressed in terms of percent of full-scale output (*% FSO*) and are easiest to explain when expressed in this manner; however, they can also be expressed in terms of a fraction of the measurand (e.g., ± 0.01 pH); sometimes they are expressed in terms of percent of reading, and in this case the allowable deviation (error) increases with increasing readings.

Accuracy refers to the allowable deviation of readings from their respective true values; in practice, such true values are established by reference to a *calibration* based on the use of a certified reference material. The proper way to specify accuracy is to use the word "within" (e.g., within ± 0.05 % FSO); otherwise the tolerance is really that of *error* (an instrument with 100% accuracy would have zero error).

Repeatability (sometimes called "reproducibility") is the ability of an instrument to reproduce output readings when the same measurand value is applied consecutively in the same direction and under the same conditions. It is expressed as the maximum difference between output readings as determined by two consecutive (preferably multipoint) calibration cycles (see Figure 2-10).

Hysteresis is the maximum difference in output at any measurand value within the (specified) range, when the value is approached first with an increas-

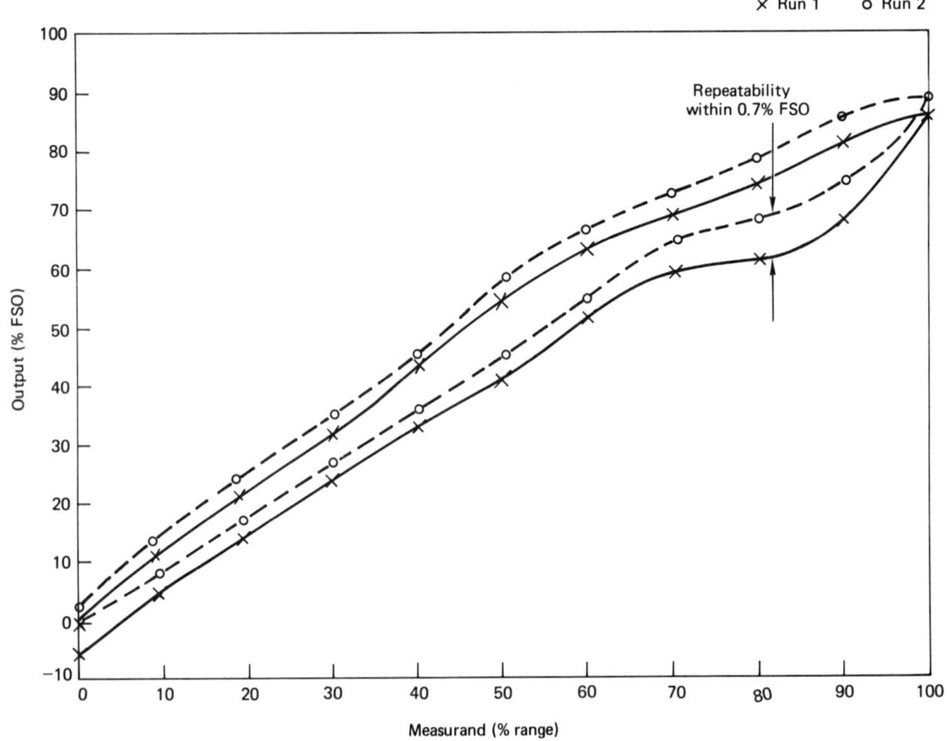

Figure 2-10. Repeatability (scale of errors 10:1).

ing and then with a decreasing measurand (see Figure 2-11). The hysteresis seen when only a portion of the range is traversed (partial-range hysteresis) is always less then the total hysteresis.

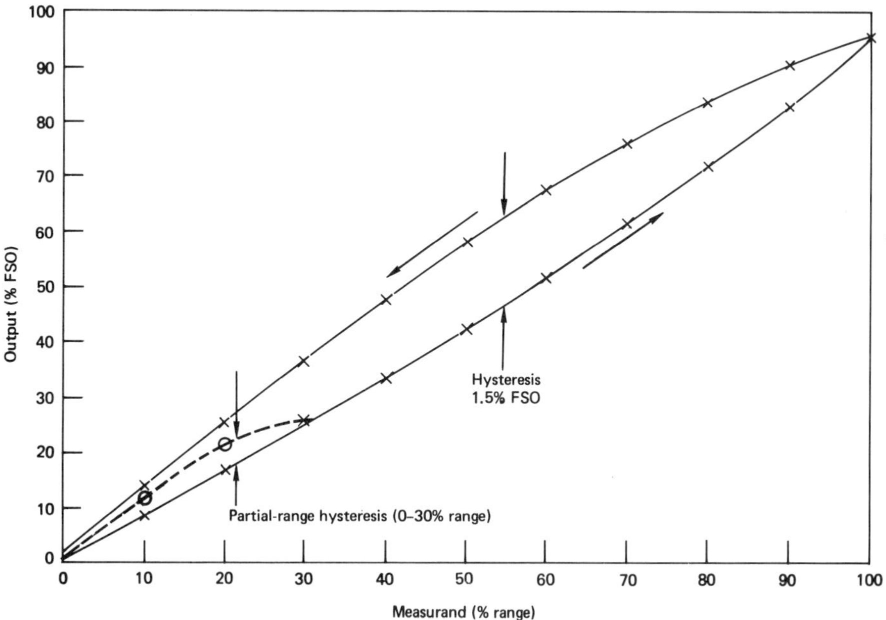

Figure 2-11. Hysteresis (scale of errors 10:1).

Linearity is the closeness of an instrument's calibration curve to a specified straight line. The tolerance applied to linearity should reflect the maximum deviation of any calibration point from the corresponding point on the specified straight line. It is really quite important to use a modifier in front of "linearity" to indicate what sort of straight line is used as reference. In general, it could be assumed that, if the type of reference line is not specified, linearity is given as independent linearity, because low tolerances for this are easiest to achieve. However, an instrument specification may show an independent linearity within ± 0.5% FSO while the terminal linearity could easily be in the order of ± 3% FSO.

Terminal linearity (see Figure 2-12) is referenced to the terminal line, a special form of *theoretical slope* (there can be other forms of theoretical slope) for which the end points are exactly 0% and 100% of both the range and the full-scale output.

Independent linearity is referenced to the "best straight line" (see

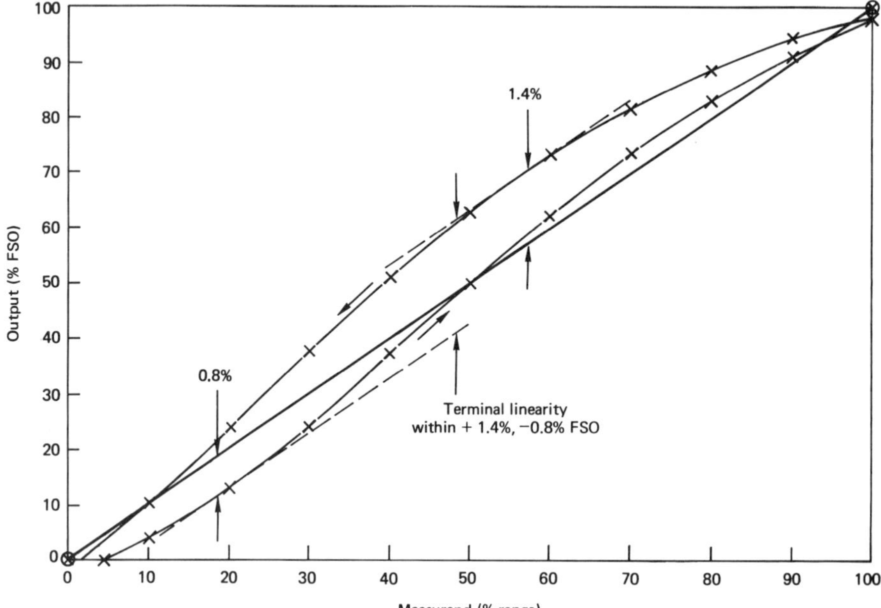

Figure 2-12. Terminal linearity (scale of errors 10:1).

Figure 2-13), a line midway between the two parallel lines closest together and enveloping all output values on a calibration curve. The best straight line can be drawn only after a calibration has been completed.

Other types of linearity include *end-point linearity,* referenced to the end-point line drawn between actual (rather than theoretical) and averaged end points, and *least-squares linearity,* which is referenced to the least-squares line, that straight line for which the sum of the squares of the residuals is minimized (this usually calls for the use of a computer). When the output-vs.-measurand characteristics of an instrument are inherently nonlinear but can be described by a theoretical curve, linearity can be replaced by *curve conformity* (conformance).

Resolution (threshold or resolving power) is the smallest change in the measurand that will result in a measurable change in output, stated in terms of the measurand. In spectrometry, spectral resolution is the smallest band of wavelengths that can be resolved (i.e., result in a measurable output change). For imaging spectrometers, both spectral and spatial resolution are important; the latter refers to the minimum dimension (on the target and at a specified distance from it) that can be resolved.

Sensitivity is the ratio of output changes to measurand changes; it estab-

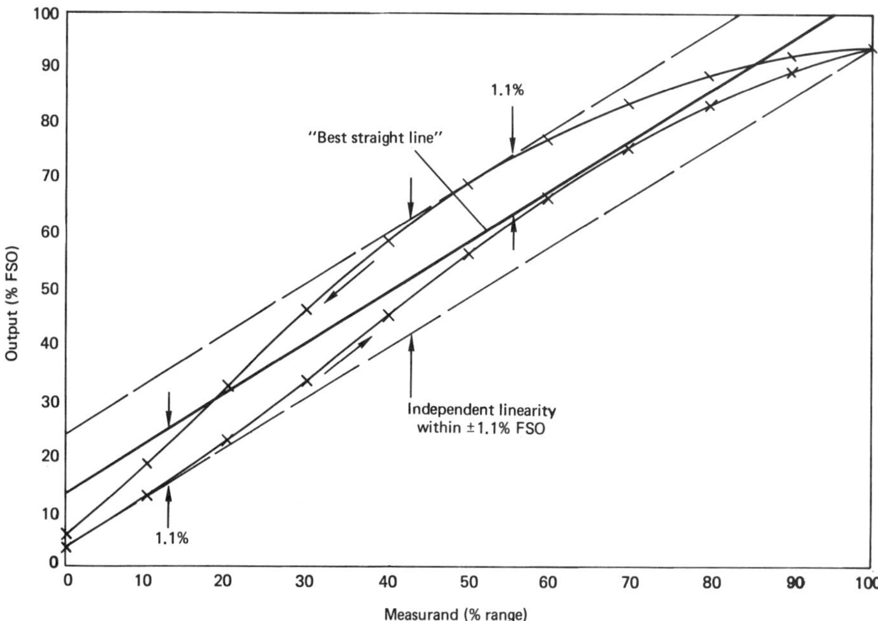

Figure 2-13. Independent linearity (scale of errors 10:1).

lishes the slope of the calibration curve. Unfortunately, many instrument specifications use "sensitivity" to specify resolution.

The *zero-measurand output* ("the zero") is the output of an instrument, under room conditions, with nominal excitation but zero measurand applied. *Zero shift* is a change of this output value over a relatively short period of time; it is characterized by a parallel displacement of the entire calibration curve. *Sensitivity shift* is a change in the slope of a calibration curve due to a change of sensitivity over a relatively short period of time. This may be days or weeks; when these periods are years, both types of shift become a reliability characteristic (*stability*).

Selectivity (*chemical selectivity*) is the ability of an instrument to respond to a specific material (e.g., gas) in the presence of (usually specified) other materials (*interfering* materials).

Transient response characteristics of an instrument are usually important to know since they define the time between bringing the instrument into contact with the sample and obtaining a valid output reading; it also defines the time, following a step change in the characteristic being sensed, before a valid output reading is obtained. When such a step change is seen by the instrument, the output will change nonlinearly toward the final value (see Figure 2-14). The

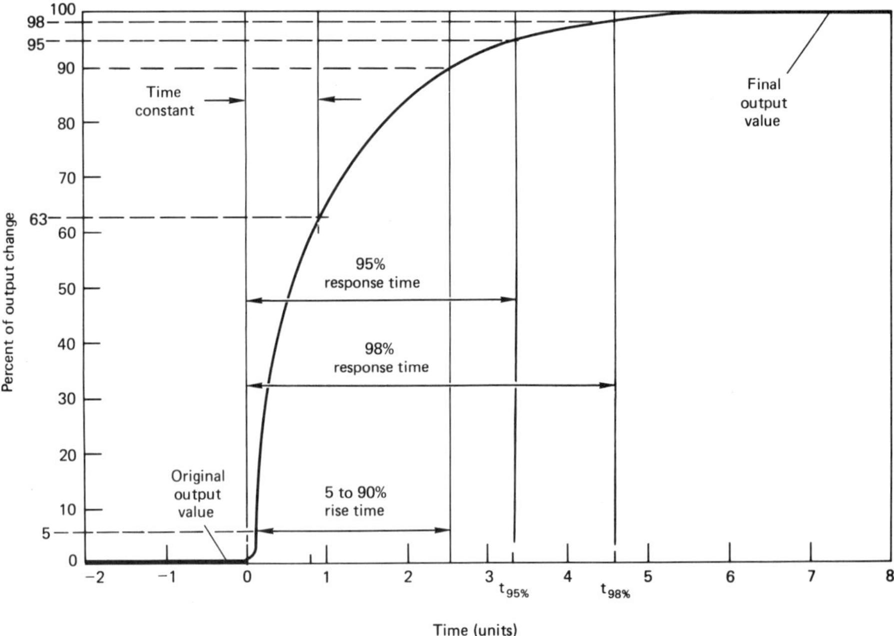

Figure 2-14. Response time, rise time, and time constant.

length of time required for the reading to rise to a specified percentage of its final value is the *response time.* The percentage should be stated in the form of a modifier of "response time," e.g., 95% response time, or 98% response time. A special term, *time constant,* has been assigned to the 63.2% response time (the symbol for time constant is τ). Another term, *rise time,* is used to state the length of time for the output to rise from a small to a large specified percentage of the final value. Those percentages are, typically, 10% and 90%.

Frequency response is the change with frequency of the output-measurand amplitude ratio within a stated range of frequencies. It is assumed that the fluctuations in measurand are sinusoidal. Frequency response should be referred to a reference frequency (see Figure 2-15) as well as to a specified measurand value. The figure shows two curves: curve A shows the response of an instrument that can be used for static as well as dynamic measurements; curve B illustrates the response of an instrument usable only for dynamic measurements.

Environmental characteristics are of two types. *Operating environmental characteristics* apply during the intended operation of the instrument, when it must operate under conditions other than those under which it was calibrated,

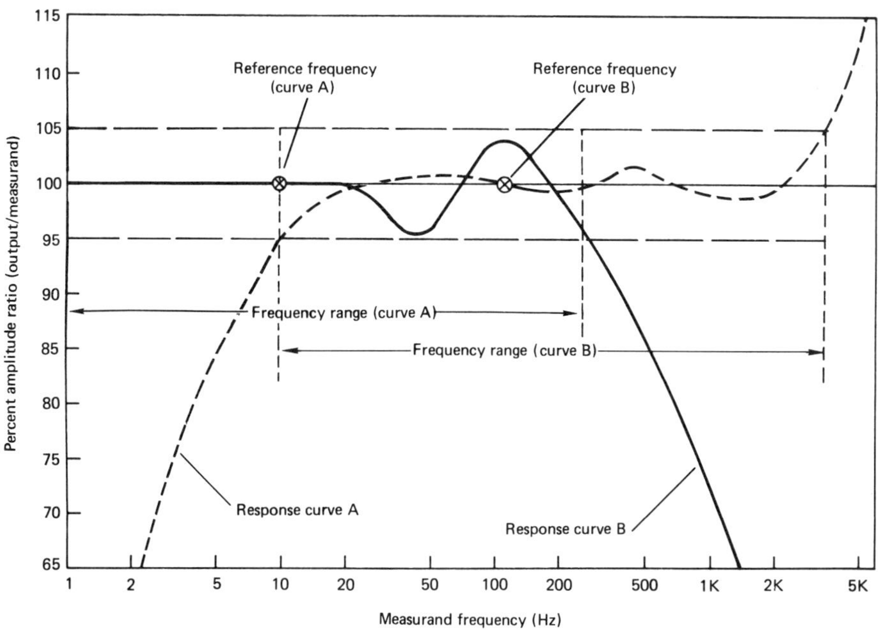

Figure 2-15. Frequency response.

for example much higher or much lower temperatures as well as shock and vibration induced by mechanical sources. Instrument specifications rarely deal with *nonoperating environmental conditions* which apply primarily during shipping, handling, installation, and storage of the instrument. Yet, some delicate instruments may be affected by vibrations induced by the vehicle that transports them, by "handling shock" (somebody drops the box holding the instrument) or by cold temperatures in the cargo compartment of aircraft. The usually specified operating environmental conditions that affect instrument performance are temperature and, in many cases, shock and vibration. When an instrument consists of a separate probe connected, by cabling, to an electronics unit, the environmental tolerances almost invariably apply only to the probe, since it can be assumed that the electronics unit is maintained at room conditions.

Temperature effects over a specified *operating temperature range* must be known for instruments that operate in a very hot or very cold ambient environment or that sense materials whose temperatures can range from cryogenic to very hot. They can be specified, with associated tolerances, as *temperature error,* the maximum change in output when the temperature is changed from room temperature to specified temperature extremes. In some cases, tolerances are separately assigned to *thermal zero shift* and *thermal sensitivity shift.* Besides operating temperature range, some specifications also show the *maximum safe temperature* that will not damage the instrument (although no performance tolerances apply at that temperature). Instruments intended for use at high or low temperatures usually incorporate electronic temperature compensation; in some cases, then, the *compensated temperature range* is stated; however, since such compensation is rarely perfect, tolerances for temperature error should still be considered. When a rise or drop in temperature is in the form of a step change, the instrument output will typically change to a value outside specified tolerances for a more or less brief period of time, before the output settles back within tolerances. This additional output error is known as *temperature gradient error,* and the time it takes the output to stabilize is one form of *recovery time* (this term is also applied to other characteristics that are subject to step changes).

Shock is a pulse of acceleration and is usually stated in terms of the maximum acceleration value (in g or m/s^2) as well as the duration of the pulse.

Vibration is vibratory acceleration and is stated in terms of maximum acceleration amplitude as well as the frequency range.

Many instruments are more sensitive to shock and vibration applied along one of their axes than along the other two axes. If this is suspected but not specifically shown in a specification, it may be wise to ask the manufacturer about the applicability of the shock and vibration specifications.

Other operating environmental effects on the behavior of an instrument during its normal operation can include excessive humidity, corrosive atmospheres or measured fluids (incompatible with the sensor material), fluids that form undesirable deposits on a sensor, slurries that abrade portions of a sensor, nuclear or ionizing radiation, and electromagnetic interference.

2.3.3 Reliability Characteristics

Instrument reliability characteristics deal with the ability of the instrument to operate within specified performance tolerances over long periods of time (years).

Operating life is the minimum length of time over which the instrument will operate (continuously or partly on, partly off), while its characteristics remain within specified tolerances.

Stability is usually shown as an increased accuracy tolerance; some long-term performance degradation can usually be expected.

Warm-up time (not really a reliability characteristic but related to stability) is the period of time, starting with the application of power to the instrument, required to assure that the instrument will perform within all specified tolerances.

Many specifications show a recommended *recalibration interval* for an instrument, and such manufacturers' guidelines should be followed meticulously.

Other reliability characteristics worthy of consideration relate to possible adverse effects of instrument failures on the system in which they are mechanically installed as well as in the system they are connected to electrically. Some of these are already controlled by existing industry or government codes. Others require a failure mode and effects analysis. Some considerations are quite simple, for example, if the instrument is electrically protected by its own fuse or circuit breaker. Others are more complex, such as those dealing with seals of probes installed into high-pressure systems (What if there is a leak in the primary seal? Can the high-pressure fluid then leak into the external housing and cause it to rupture, thus posing a hazard to adjacent equipment as well as to nearby personnel?).

Generally it is wise to assess possible failure modes, even if they have a low probability of occurring, and then take steps to build appropriate protection into the instrument as well as into the system.

2.4 INSTRUMENT SELECTION CRITERIA

The selection of a particular instrument design is based primarily on an analysis requirement. This requirement is sometimes established by the person responsible for the selection; much more frequently it is established by someone else, typically a project engineer or the cognizant engineer of a subsystem. Usually the data system (including power supplies), with which the instrument must interface and operate, either exists or has already been designed. There can also be other constraints on instrument selection. There may be a project policy to use only parts (including instruments or electronic parts within a instrument) that are on an "approved list" or that have previously passed a qualification test for specific applications; the purchasing department may have a list of approved vendors, based on prior procurement and quality control experiences; or, regulations may prohibit buying instruments from another country unless it can be proven that a manufacturer there is the only possible source. These are examples, and there are usually sound reasons for such policies. It should be evident, then, that instrument selection is often an iterative process and that there can be times when analysis requirements have to be negotiated with the originator.

The listing of guidelines, below, is meant to be fairly complete. For some applications, a number of these considerations can be omitted. For others, additional factors may have to be considered. Some of the entries tend to imply possible incompatibilities. For example, if the analysis requirement calls for an instrument (digital) output data rate of 1 Mb/s and the data system can only handle up to 100 kb/s, the analysis requirement will probably have to be negotiated downward. Cost and availability can affect analysis requirements similarly.

2.4.1 Analysis Considerations

1. What is the real purpose (the objective) of the analysis?
2. What is the instrument intended to analyze?
3. Is the analysis to be quantitative or qualitative? Single-component or multicomponent?
4. What range of values will be displayed in final data?
5. Is hysteresis a concern (will the property to be analyzed only increase, or only decrease, or both)?
6. With what accuracy must the analysis be presented in final data?
7. What are the dynamic characteristics of the property to be analyzed (fluctuation frequency range, step changes)?
8. What frequency response or transient response must be visible in final data?
9. If a fluid is being measured, what are its physical and chemical characteristics (other than the property being analyzed)?
10. Where and how will the instrument be used and installed?
11. In what manner, and to what extent, is it permissible for the instrument to modify the property to be analyzed, during analysis?
12. What ambient environmental conditions will the instrument be exposed to?

2.4.2 Data System and Power Supply Considerations

1. What is the general nature of the data system (e.g., radio telemetry, hardwired telemetry, individual direct display)?
2. Is the data system inherently analog or digital?
3. What is the nature of the major elements in the data system:
   a. Signal conditioning including amplification (linear or logarithmic), analog-to-digital conversion, multiplexing, pre-transmission buffering?
   b. Data transmission link (if any)?
   c. Data processing, data storage?
   d. Computer interface?
   e. Data display?

4. What are the accuracy and frequency response characteristics of the end-to-end data system exclusive of those of the instrument?
5. What form of instrument output will the data system accept with minimum additional signal conditioning?
6. What load impedance will be seen by the instrument?
7. Is frequency filtering or amplitude limiting of instrument output required, and can the data system handle this?
8. What instrument excitation voltage is most readily available?
9. How much current may the instrument draw from the excitation supply?
10. Are special instrument-related checking functions (e.g., "ready" check, electrical calibration check) required by the data system and does the data system provide circuitry for these?
11. Which of the above functions are usually incorporated within the instrument? Which additional functions can optionally be incorporated and at what cost?

2.4.3 Instrument Design Criteria

In the step-by-step process offered by this chapter, criteria for instrument design, and, hence, selection will be based primarily on the lists of considerations, above. However, as was stated at the outset, cost and availability factors, as well as policies governing procurement, may influence design decisions and can translate into analysis requirement changes.

1. What constraints are imposed on instrument mass, configuration, excitation, and power consumption?
2. What are the instrument output (or built-in display) requirements?
3. Which operating principle is most suitable?
4. What accuracy and other performance characteristics must the instrument provide: static? dynamic? environmental (operating)? environmental (nonoperating)?
5. What operating or cycling life is required?
6. If a fluid is to be measured, what effects of the measured fluid on the instrument must be considered?
7. Will the instrument affect the analyzed property to the extent that erroneous data will be obtained?
8. What constraints are imposed on the instrument by any applicable governmental standards or industry codes?

9. What are the failure modes of the instrument? What hazards would a failure present to the system in which it is installed, to adjacent components or systems, to the data system, to the power supply, to the area in which the instrument operates, or to personnel working in that area?
10. What is the lowest level of technical competence of any and all personnel expected to handle, install, and use the instrument? What human-engineering requirements should be incorporated in the instrument design?
11. What test (including calibration) methods will be used to verify performance? What tests will be performed by the manufacturer and what tests will be run by the user? Are those tests adequate? Are test methods correct? Is test equipment appropriate? Are test methods simple and well established?

2.4.4 Availability Factors

1. Is an instrument that fulfills all requirements available "off-the-shelf"?
2. If the answer to the above question is no, the following should be considered:
   a. Will minor redesign of an existing instrument be sufficient or will a major development effort be required?
   b. How many instruments of identical design will be procured at this time and possibly at future times?
   c. What manufacturer has produced an instrument similar to the required item?
   d. What past experience exists in dealing with a proposed manufacturer?
   e. Can the instrument(s) be delivered in time to meet installation or usage schedules?

2.4.5 Cost Factors

1. Is the quoted cost of the instrument compatible with the analysis function it will provide?
2. What additional costs will be incurred by required instrument testing, periodic recalibration, handling, and installation?
3. Which requirement imposed on the instrument is the major cost driver?
4. What relatively minor compromises in requirements could lead to substantial savings?
5. What modifications to the data system or to the power supplies could lead toward reduced costs of a number of different instruments used in that system, and what cost trade-offs would be involved?

2.4.6 Location of Processor ("Smart" Instruments)

Considerations of data system characteristics were covered in Section 2.4.2. An additional consideration for selection of an instrument is the location of the processor that performs at least some of the data processing as well as some instrument control functions. The continuing development of microprocessors has made it possible to package such a processor integrally with the instrument; that is, it is contained within the instrument housing. The capabilities of such built-in processors vary over a wide range. Some fairly simple instruments are available with a built-in processor that performs such operations as analog-to-digital conversion, storage of the calibration curve and conversion factors, and output of data in engineering units in digital form. Additional capabilities can be internal buffering (storage) of data to provide a data dump when polled by the data system; decoding of command sequences and their conversion into instrument operating mode changes; responding to an instrument's own observation and changing an operating mode (e.g., amplifier gain) and sampling rate; combining data words with time tags obtained from a clock signal, with error correction codes, with data about internal operational states, and with headers that the data system accepts as the beginning of a data sequence (frame).

Instruments that incorporate such microprocessors are often called "smart" or "intelligent" instruments. Making a choice between a "smart" and a "normal" instrument involves primarily trade-offs of cost vs. data system capability vs. convenience, assuming here that more than one instrument is connected to the same data system. For example, multiplexing standardized instrument outputs into a single analog-to-digital converter (ADC) is probably cheaper than paying for an ADC for each instrument; on the other hand, multiplexing (formatting) digital data may facilitate programmability. Or, if the existing data system already provides for programmable decalibration and engineering-unit conversion, there is really no need to have these functions provided by a microprocessor in the instrument. On the other hand, if the data system is new and was specifically designed for these functions to be performed by each instrument, the overall cost may be the same, and the cost of expanding the data system to handle additional instruments may even be less.

3

ELECTROMETRIC ANALYSIS INSTRUMENTS

The operation of electrometric analyzers is based on the electrical characteristics of an *electrochemical cell* in which specific reactions occur. The electrical characteristics are measured individually or in combination and comprise current, voltage, and resistance (or its reciprocal, conductance); additionally, the variation of one or more electrical characteristics with time can provide information about chemical properties and composition. The field in which *electrometric methods of analysis* are used is known as *electroanalytic chemistry.*

The primary application of electrometric analysis instruments is in the determinations of concentrations of solutions. Concentration is now most commonly expressed in terms of *molarity,* the number of gram-molecular weights of a substance present (dissolved) in 1 liter of solution; molarity is indicated by the symbol *M,* preceded by a number to show solute concentration (e.g., 2.4×10^{-5} *M*). Other applications include determinations of the acidity or alkalinity (*pH*) of a liquid, of the magnitude of the oxidation or reduction potential, of the presence/absence of substances in liquids, and of the relative amount of a specified gas in a gas mixture referred to a reference gas mixture containing a known amount of the specified gas.

Electrode potentials are composed of the potentials produced by two half-cells, in combination. One half-cell contains the electrode of interest; the other half-cell is a standard hydrogen electrode. The potential E of the electrode of interest is then given by the *Nernst equation,* which, at a temperature of 25°C, can be expressed in the form

$$E = E_0 - \frac{0.05915}{n} \log \frac{a_{red}}{a_{ox}}$$

where E_0 is the standard-electrode potential, n the valence change (number of electrons transferred in the electrode reaction), and a_{red} and a_{ox} are the activities of the reduced and oxidized forms, respectively, of the reactants (of the electrode action).

3.1 CONDUCTIVITY SENSORS

Conductivity measurements are made primarily to determine the concentration of a solution or to determine the relative amount of a salt in an aqueous solution. The principle employed is that of *electrolytic conduction,* in which the charge carriers are provided by ionization. When a current flows through a volume of a solution, the soluble inorganic compounds in the solution will partially or completely separate into *cations* (positively charged ions) and *anions* (negatively charged ions). For example, NaCl forms Na^+ and Cl^- ions, $AgNO_3$ forms Ag^+ and NO_3^- ions, and $BaCO_3$ forms Ba^{2+} and CO_3^{2-} ions. Generally, hydrogen and metal ions are cations, and negative radicals and nonmetal ions are anions. The ionizing current originates from two electrodes, a cathode and an anode. The cations migrate to the cathode, at which they combine with the electrons of the current source to form hydrogen or metal atoms. The anions migrate to the anode, at which they form neutral atoms or molecules and liberate their electrons, which then flow to the current source.

When a potential is applied across two electrodes which are immersed in a solution, the current flowing through the circuit will be a function of the applied voltage and the resistance of the solution. This resistance, in turn, is a function of the nature of the solvent (water, and hence a constant factor, in the case of aqueous solutions), of the number of ions present, and of the ion mobility. Conductance (G) is the reciprocal of resistance (R) or $G = 1/R$. The (electrolytic) conductance of a solution, then, is proportional to the number of ions and to ion mobility. Conductance is expressed in *siemens* (S); it was formerly expressed in *reciprocal ohms* (*mho*). The parameter that characterizes solution concentration is (electrolytic) *conductivity* (γ), expressed in *S/m* (submultiples such as μS/cm are commonly used).

The conductivity measurement is given by the measurement of the conductance of the liquid column between two electrodes. Assuming two parallel plate electrodes, each having the same area A, separated by the distance L, with a uniform separating volume, the conductivity, $\gamma = G \cdot L/A$. When the conductivity is expressed in siemens, the electrode area in cm^2, and the separation distance in cm, conductivity will be expressed in S/cm, its most common unit. The ratio L/A is a constant for a given electrode configuration and is known as *sensor constant* (also called cell constant or electrode constant), K_s; hence, $\gamma = GK_s$.

The type of excitation voltage applied to conductivity sensors is critical to its performance. If pure dc were applied, a resulting plot of current flow vs. time would show, first, a short spike due to the capacitance of the sensor and its connecting leads, and next, a short period where the current is purely dependent on conductance; this is followed by a long period during which electrolysis at the probe surface causes the current to decrease until a minimum current is reached when the sensor is polarized. For this reason, ac sensor excitation is used. When the waveshape is sinusoidal, an average of the four states described above would be obtained. The only current of interest is the short plateau following the initial spike. This points to the requirement for a square-wave excitation voltage; it should have a slow rise time so that the capacitance-caused spike is minimized; it should also have a fairly high frequency so that polarity is reversed before electrolysis becomes significant. The negative error still present due to averaging can be minimized by employing a peak-detector circuit. The effects of lead resistance can be minimized by using a four-wire connection to the associated bridge circuit, as is done for resistance thermometers. Lead-resistance effects as well as polarization are further reduced by using separate voltage and current electrodes.

Figure 3-1 illustrates a conductivity measuring circuit. The square-wave excitation is applied to the bridge circuit in which a potentiometer allows manual balancing and into which the electrodes as well as thermal compensation elements are connected. The signal due to current flowing across the electrodes is amplified and rectified. The time constant of the capacitor (C_1) in conjunction with the associated resistors is such that it is charged by amplifier (A_2) to the peak value seen; this avoids an output averaging with the lower currents due to electrolysis on the electrodes. The final output voltage is then displayed on a meter.

Configurations employing two parallel plates are no longer used; they are shown in Figure 3-1 primarily for illustrative purposes and referred to in the text to explain cell constant fundamentals. Modern conductivity sensors are available in a variety of configurations. The relationships for their electrode geometries are more complex than for parallel plates; these relationships are now well developed and cell constants have been determined. The most common examples for modern electrode geometries are parallel rods and concentric annular electrodes. Typical electrode materials are nickel,

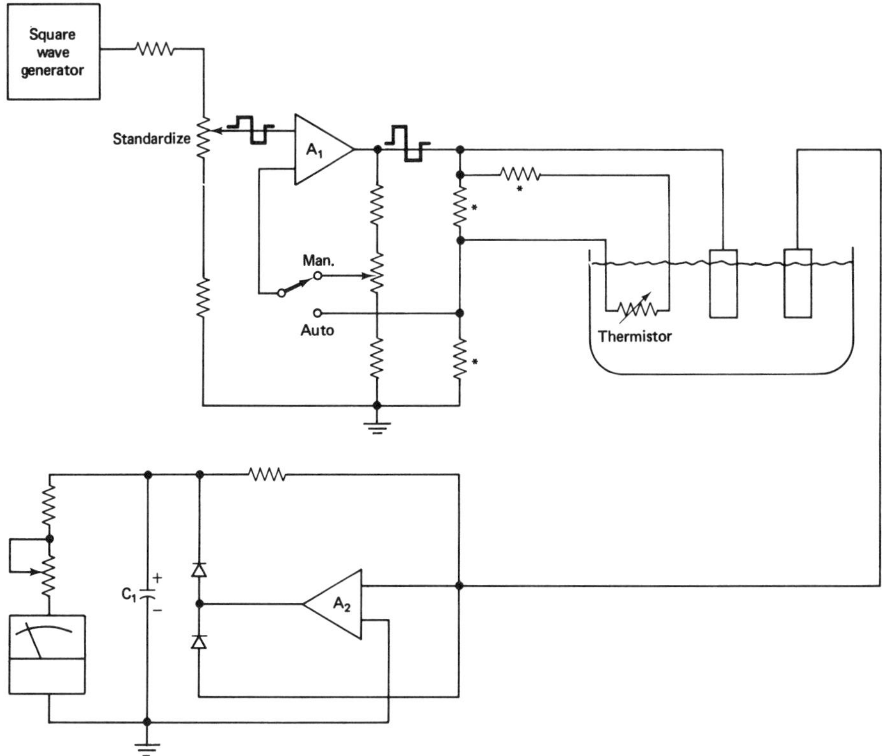

Figure 3-1. Typical conductivity measuring circuit; the effect on gain of A1 is characterized by the series-parallel resistances (*) associated with the thermistor and the combination compensates for fluid temperature variations; sensor elements are shown separately in this schematic representation. (Courtesy of Rosemount Analytical, Inc.)

carbon, stainless steel, and ferrous-nickel alloys; gold plating is sometimes used. Insulating materials include polyvinylchloride, epoxies, silicone rubbers, and high-temperature corrosion-resistant (proprietary) plastics. Since measured-fluid temperature has severe effects on the conductivity measurements, metal-film, metal-wire, or thermistor thermometers are usually incorporated in conductivity sensors; some designs contain provisions for adding such a thermometer, or the sensor specification makes the installation of a thermometer in close proximity mandatory.

Figure 3-2 shows some insertion-type conductivity probes (screw-in cells). They are mounted into a threaded boss. Housings are stainless steel or plastic. Other conductivity probe designs include sanitary cells mounted by

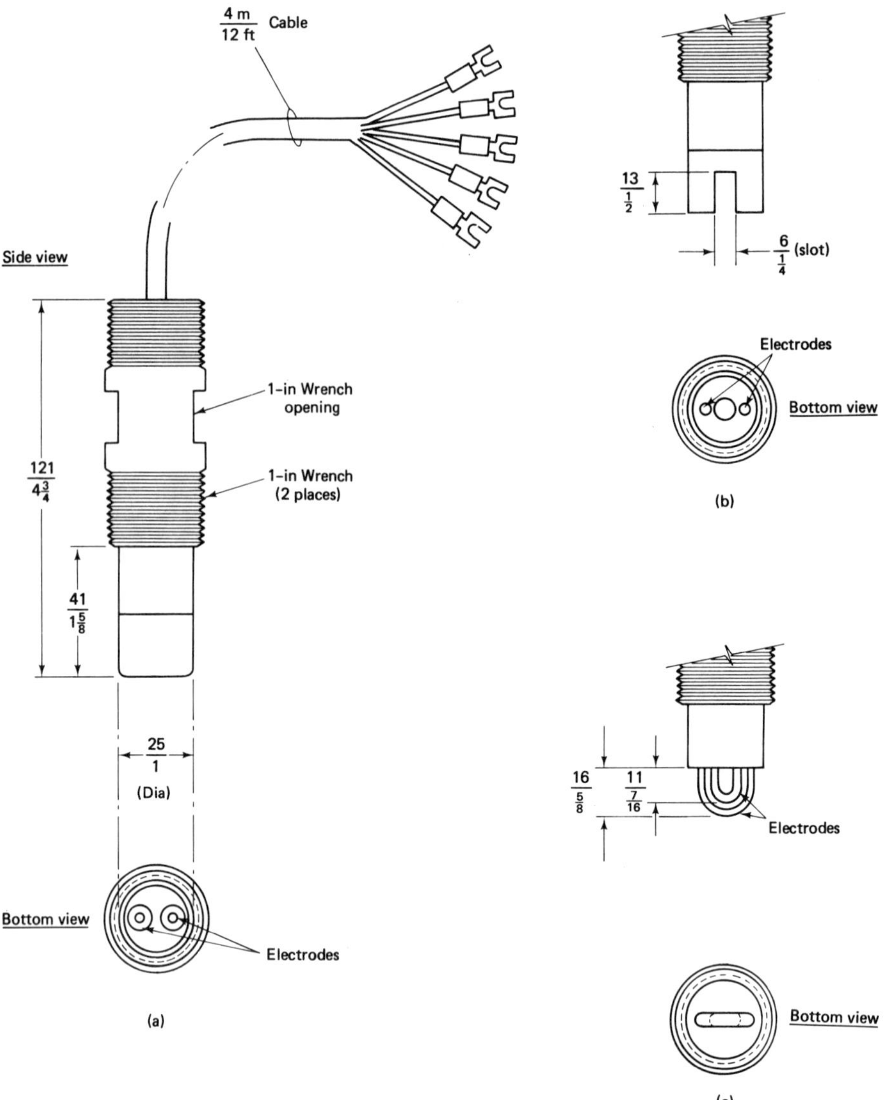

Figure 3-2. Typical insertion-type conductivity probe and electrode configurations and their cell constants, with dimensions in mm/inches: (a) $K_s = 2.0$; (b) $K_s = 1.0$; (c) $K_s = 0.1$. (Courtesy of Rosemount Analytical, Inc.)

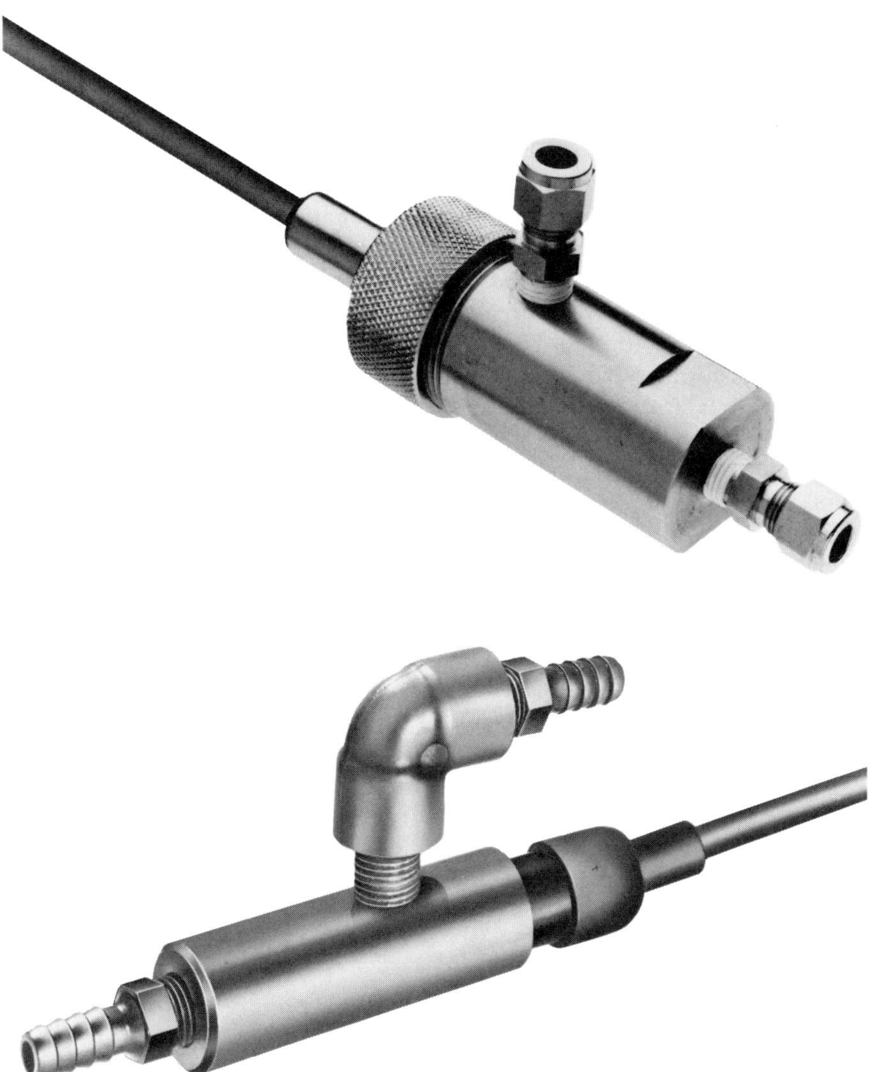

Figure 3-3. Flow-through conductivity cells: (a) stainless-steel flow cell; (b) plastic flow cell. (Courtesy of Rosemount Analytical, Inc.)

special clamps and dip cells which are not mounted but hand-held. Ball-valve and gate-valve cells are available for installations in which the cell must be able to be removed from the line while the line is under pressure. Figure 3-3 shows two types of flow-through conductivity cells. The stainless-steel version is rated at 100 psi at temperatures up to 100°C. The plastic version has a PVDC body and is rated for 100 psi at 25°C, derated to 8.3 psi at 100°C. Both versions are available with cell constants of either 0.01/cm or 0.1/cm. Sensors containing four, rather than two, electrodes connected in an appropriate circuit are available for measurements of very high concentrations.

In some applications, the exposure of electrodes to measured fluids is not advisable, due to the chemical or physical characteristics of the fluid (e.g., brine, two-phase liquids, abrasive or fibrous slurries). This problem can be overcome by using an electrodeless (*toroidal*) conductivity sensor (see Figure 3-4). Such sensors are available in submersible, insertion, or flanged flow-through designs. A toroidal conductivity sensor is, in essence, a variable-coupling transformer. Two toroidal windings, spaced a small distance apart in a single housing, are encapsulated and sealed with good physical and chemical protection from the measured fluid, thus completely isolated from the fluid. One toroid, excited by ac, usually at audio frequencies, acts as the primary (input) winding. The other toroid acts as the secondary (output) winding. The conductive measured fluid, itself, provides the variable coupling between the windings. The coupling, and, hence, the output signal is proportional to conductivity. Toroidal conductivity sensors provide measuring ranges between 0–100 μS/cm and 2 S/cm. The measurement is not affected by flow rate or flow direction. A temperature sensor is incorporated into the conductivity cells.

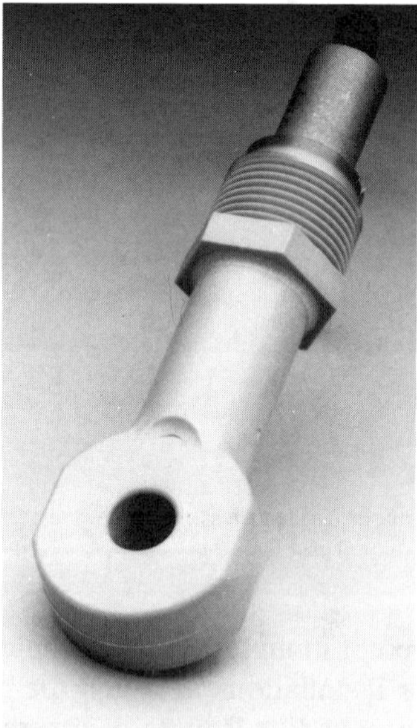

Figure 3-4. Toroidal conductivity sensor. (Courtesy of Rosemount Analytical, Inc.)

Most conductivity measurement systems consist of a sensor connected by a cable to the associated electronics and read-out unit. However, compact, hand-held conductivity meters, with a probe that is dipped into the measured fluid, are also available. The meter shown in Figure 3-5 is powered by a 9 V battery.

Figure 3-5. Portable conductivity meter. (Courtesy of Pathfinder Instruments.)

The most common applications of conductivity measuring systems are in concentration measurement of solutions (the measurement can then also be used for dilution control) and in measurements of the amount of total dissolved solids in such liquids as water. In fact, water analysis is one of the prime applications of conductivity sensors.

For single-component analyses of concentration, the output readings are correlated with concentration by means of graphs, charts, and records of previously obtained readings. The determination of concentrations of a specific component in a mixed solution is more difficult. Generally, such a measurement can be obtained if the concentrations of the nonmeasured components remain reasonably constant, or when the measured component has a much higher conductivity than the other components. For example, most acids and bases are much more conductive than their salts, since hydrogen and hydroxyl ions have a very high mobility. Conductivity sensors

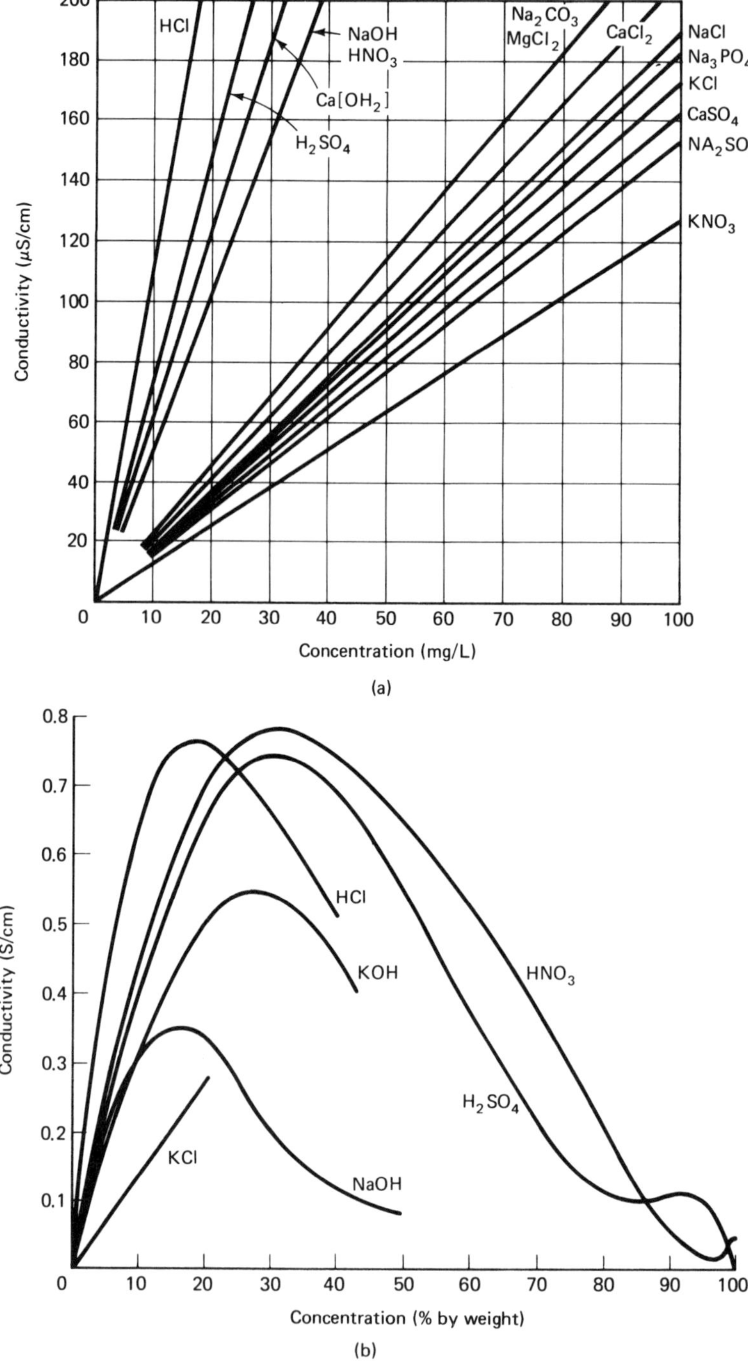

Figure 3-6. Conductivity as a function of concentration for several solutions, at 18 °C: (a) relationship is linear for relatively low concentrations; (b) relationship becomes nonlinear and slope reversal often occurs for very high concentrations.

are also used for the detection of leaks and spills into liquids whose conductivity would be affected significantly by such events, and for salinity measurement.

Due to mass increases of ions with increasing concentrations, the conductivity vs. concentration relationship becomes increasingly nonlinear, in many cases actually reaching a peak followed by a slope reversal. This characteristic is illustrated in Figure 3-6 (note that the conductivity units are μS/cm in Figure 3-6a, whereas they are S/cm in Figure 3-6b). However, the curves will be repeatable; hence, it is still possible to determine very high concentrations even when slope reversal occurs, as long as a separate check is made to find out whether the higher or the lower concentration is being measured.

Conductivity sensors can also be used, in conjunction with elements that cause a specific chemical reaction resulting in conductivity changes, for analyses not related directly to conductivity. An example of this is a *dissolved-oxygen analyzer* using a cartridge filled with thallium shavings and two conductivity sensors. Thallium is not affected by oxygen-free water, but oxidizes in water containing dissolved oxygen. The resultant thallium oxide combines with the water to form thallium hydroxide (TlOH), which is easily soluble in water and then increases the conductivity of the water sample. In one instrument using this principle, the sample is made to flow, first, through a cation and anion exchanger that reduces the conductivity of the sample to below 1.0 μS/cm (to establish a low reference value), next, through a conductivity sensor, then through the thallium reactor, and finally through a second conductivity sensor. The residual conductivity, as measured by the first sensor, is subtracted from the conductivity sensed by the second sensor, and the differential value is then indicative only of the amount of TlOH created by dissolved oxygen in the sample. The quantity of dissolved oxygen can then be calculated from this value.

3.2 pH SENSORS

The *pH* of a solution is a measure of its hydrogen ion activity and is indicative of the *acidity* or *alkinity* of the solution. pH is expressed in numbers on a scale of 0 to 14. The number 7, at the midpoint of this scale, represents a neutral solution, the pH of pure water. pH values decreasing from 7 to 0 indicate increasing acidity, whereas pH values increasing from 7 to 14 indicate increasing alkalinity. The pH number represents the negative logarithm of the hydrogen ion activity, or $\mathrm{pH} = -\log_{10} a_{\mathrm{H}^+}$, where a_{H^+} is the hydrogen ion activity. The activity of the H^+ ions (or *hydronium* ions, H_3O^+) increases with increasing acidity. As the activity of the hydrogen ion increases in a solution, the activity of the negative ions (OH^-, Cl^-, etc.) decreases accordingly. The reverse (increasing negative-ion activity, decreasing H^+ activity) is true for increasing alkalinity. The logarithmic relationship means that, for each pH value decreasing from 7, the activity of the hydrogen ion increases by one order of magnitude, whereas it decreases by an order of magnitude for each pH value increasing from 7 to 14.

The pH measuring circuit consists of two electrodes immersed into the measured fluid (solution) and a voltmeter connected across the two electrodes. Of the two electrodes, one is the *pH sensing electrode;* the other is the *reference electrode.* A basic pH electrode is the *glass electrode* (Figure 3-7a). The principle on which the function of this electrode is based originated in the discovery, by the German chemist Fritz Haber in 1901, of voltage changes, seen at certain glass surfaces, that varied in a regular manner with changes in solution acidity. As shown in the illustration, the electrode assembly consists of a glass body with a cap; the electrode wire, which may be shielded, and which is connected to the insulated (typically shielded coaxial-cable) external connecting lead, extends through the center of the electrode body and terminates in an internal reference electrode. The bulb at the bottom of the electrode is filled with a solution characterized by constant pH and reference ion activity. The reference electrode is immersed in this solution, which wets the inside surface of a membrane of specially formulated glass. The outside surface of this pH-sensitive membrane is wetted by the solution to be measured.

There are three interfaces in and on this electrode at which electrical potentials are developed. The potential at the reference electrode/constant-pH solution interface remains constant. The potential at the constant-pH solution/glass membrane interface also remains constant. The potential at the glass membrane/measured solution interface, however, is proportional to pH. The nominal voltage produced is about 60 mV per pH unit.

Figure 3-7b illustrates the generalized elements of a reference electrode. A wire electrode is immersed in a solution which is in contact with a salt solution (bridging solution) that comes in contact with the measured fluid (measured solution). Again, there are three interfaces at which electrical potentials are developed: the reference-metal-wire/reference-metal-ion solution interface, the reference-ion solution/salt interface, and the salt solution/measured solution interface. All three of these potentials remain constant. A frequently used reference electrode is the *silver–silver chloride* electrode. The reference electrode is a silver wire, coated with solid AgCl. The salt solution is KCl. Another reference electrode, also used quite commonly, is the *calomel* electrode. Calomel is a common name for mercurous chloride (Hg_2Cl_2). In this type of reference electrode a metal wire (platinum) is in contact with solid mercury which, in turn, is in contact with solid

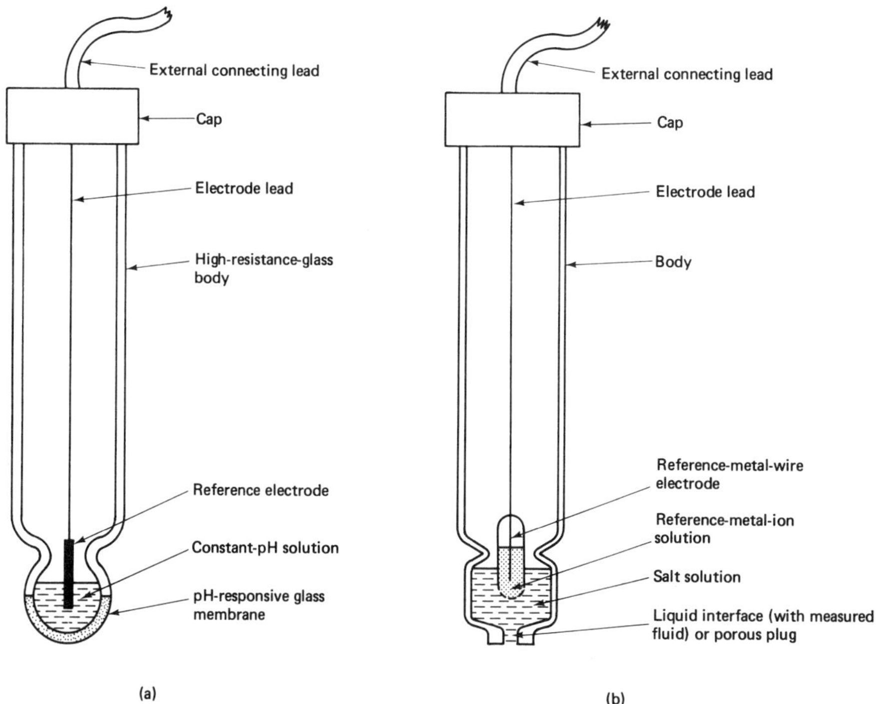

Figure 3-7. Schematic representation of pH sensing electrodes: (a) pH electrode; (b) reference electrode.

Hg_2Cl_2, and the calomel, in turn, is in chemical contact with a (typically saturated) solution of potassium chloride (KCl), which then acts as the bridging solution. The junction between the bridging and the measured solution is either in the form of a thin capillary opening (liquid junction) or in the form of a porous plug. Other types of reference electrodes have also been developed.

The associated voltmeter must have an input impedance that is much higher than the impedance of the electrodes, which can be up to 500 MΩ. Input stages employing a field-effect transistor (FET) can meet this requirement and are now often used in pH meters. The voltage provided by the electrodes is affected by measured-fluid temperatures such that the thermal sensitivity shift of the measuring system is about 0.197 mV/pH/°C. When the measured-fluid temperature is known and constant, temperature compensation can be achieved by manual adjustment of the meter. When the measured-fluid temperature is expected to fluctuate, a temperature sensor is either immersed separately into the fluid and connected to compensating circuitry, or incorporated in one of the electrodes and similarly connected to compensate the electrode potentials for temperature effects.

The nominal output of a pH electrode is 59.1 mV/pH, at 25°C. The actual output of any one electrode can vary from this value. Hence, a sensitivity (slope) adjustment is provided (separate from that used to adjust for sensitivity shift due to measured-fluid temperature) on most pH meters. This is known as the "standardize" adjustment. It is intended to be used in conjunction with use of a standard solution, of accurately known pH, into which the electrodes (or combination electrode) are immersed; the "standardize" knob is then rotated until the meter reading corresponds to the pH of the standard solution.

Most pH sensors in current use employ *combination electrodes,* a sensing electrode and a reference electrode mounted side-by-side (sometimes concentrically) in a single body, usually made of either a polymer or glass, sometimes

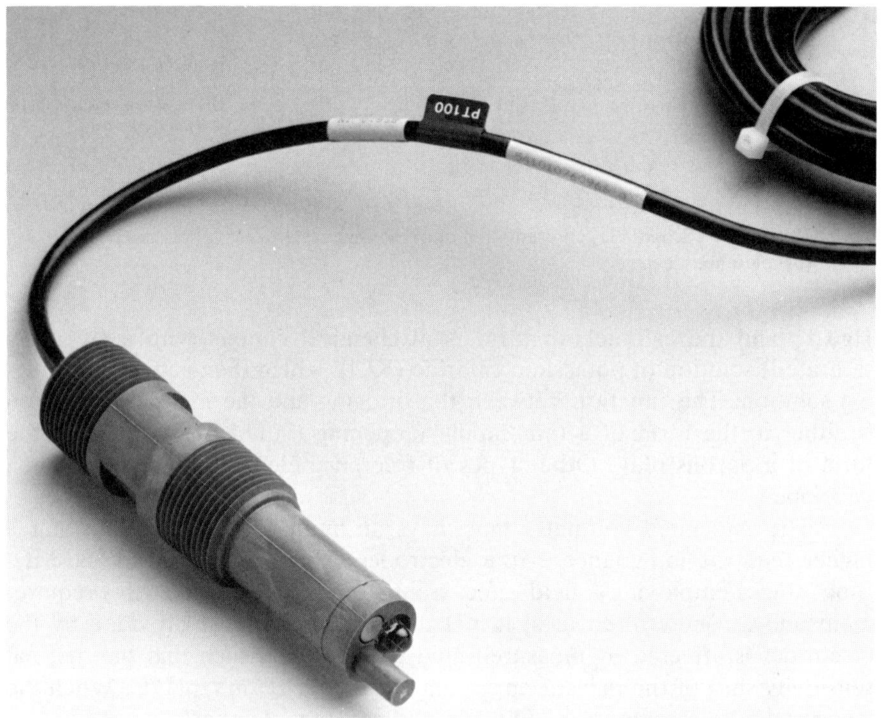

Figure 3-8. pH probe with integral platinum resistance thermometer and 6.5 m (20 ft) cable. (Courtesy of Rosemount Analytical, Inc.)

of stainless steel. The internal reference is Ag/AgCl or calomel. Some designs use a gel rather than a liquid as internal reference. In *double junction* sensors, the Ag/AgCl is contained in an inner chamber and a chemically compatible reference solution is contained in an outer chamber; such sensors are useful when the measured solution contains organic compounds, heavy metals, or other compounds that interact with silver.

Usually, pH probes are either of the dip-cell type (hand-held) or threaded for submersion mounting, in-line mounting into a tee, or into a tee of special design that splits the stream into a flow-reduced side stream. Figure 3-8 shows a typical pH probe with threads for insertion mounting; the probe incorporates a 100 ohm platinum resistance thermometer for simultaneous temperature sensing; it uses a double-junction reference cell and has a polyvinylidene fluoride body and integral cable. The same manufacturer offers a similar probe design with a single reference junction, a recessed probe tip, and an electrical connector.

pH measurement systems typically comprise a probe connected by a cable to an electronics/read-out unit. Figure 3-9 shows a bench set-up with a pH probe, on an adjustable arm, submersed into a sample beaker. The programmable microprocessor-based electronics/read-out unit can display pH, mV and temperature (the probe incorporates a temperature sensor); this unit includes a microprinter. Self-contained hand-held pH meters are also avail-

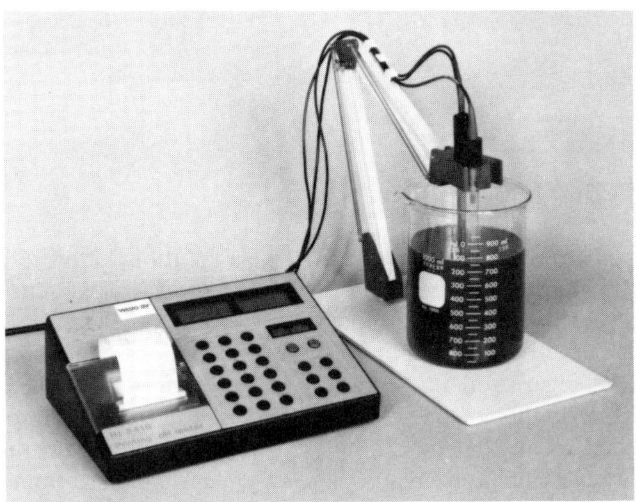

Figure 3-9. pH measuring system in laboratory set-up. (Courtesy of Presto-Tek Corp.)

able. Figure 3-10 shows a pH meter and a combination pH-conductivity meter. Both are battery-powered.

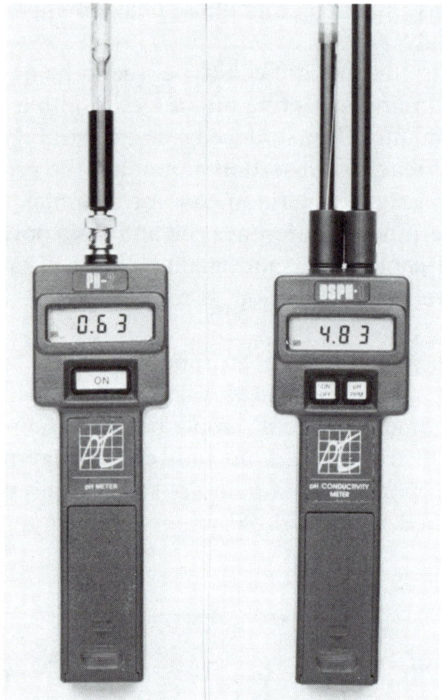

Figure 3-10. Hand-held, self-contained pH meters: (a) for pH only; (b) for pH and conductivity. (Courtesy of Pathfinder Instruments.)

3.3 ORP (REDOX) SENSORS

The *oxidation-reduction potential* (*ORP*) of a solution, often referred to as "redox," is measured by the same devices as used for pH sensing, except that the pH sensing electrode is replaced by a platinum electrode (sometimes gold, or gold or silver on a platinum base). Oxidation-reduction potential is related to the logarithm of the ratio of the activities of the oxidized and reduced states of the ions. The same type of reference electrode as prescribed for pH measurement is also used in conjunction with the platinum electrode for ORP measurements. The two electrodes are immersed adjacently; alternatively, both electrodes are packaged into a single combination electrode. The electrode potentials are read out directly in millivolts. A positive ORP indicates that the solution contains a significant oxidizing agent. A negative ORP indicates that the solution contains a strong reducing agent. A solution that is neither reducing nor oxidizing will have zero ORP. Various electrode constructions are used, including those employing a (usually platinum) wire protruding from, and fused into, an electrode body with a glass tip, or a section of platinum foil, or platinum deposited on a substrate. Some measurement systems allow for the measurement of pH as well as ORP by providing a mV as well as a pH scale, and by either switching between a pH and an ORP electrode, or by providing a suitable electrode combination in a single body.

3.4 SPECIFIC ION SENSORS

In addition to the ORP-specific platinum electrodes and the hydrogen-ion-specific pH electrodes, electrodes have been developed that are specific to (respond specifically to the activity of) one of a considerable number of other ions (see Table 3-1).

TABLE 3-1. Ions for Which Specific-Ion Electrodes Are Commonly Available

Ammonia, NH_3 (ammonium, NH_4^+)	Cupric, Cu^{2+}	Potassium, K^+
Arsenic, As^{5+}	Cyanide, CN^-	Silver, Ag^+
Bromide, Br^-	Fluoride, F^-	Sodium, Na^+
Cadmium, Cd^{2+}	Iodide, I^-	Sulfate, SO_4^{2-}
Calcium, Ca^{2+}	Lead, Pb^{2+}	Sulfide, S^{2-}
Carbonate, CO_3^{2-}	Mercuric, Hg^{2+}	Sulfur dioxide (sulfite, SO_3^{2-})
Chloride, Cl^-	Nitrate, NO_3^-	Thiocyanate, SCN^-
Chlorine, Cl_2 (hypochlorite, ClO_3^-	Nitrogen oxide (nitrite, NO_2^-)	Zinc, Zn^{2+}
Chromium, Cr^{6+}	Perchlorate, ClO_4^-	
	Phosphoric, P^{5+}	

Such *specific-ion electrodes* (*ion-selective electrodes, ISE*) can be cation-specific (respond to positively charged ions) or anion-specific (respond to negatively charged ions), and can respond to monovalent or divalent ions. Specific-ion electrodes are also always used in conjunction with a reference electrode which completes the electrical circuit at the voltmeter input. The electrode potential developed over the nominal ion-activity measuring range may be symmetrically or asymmetrically bidirectional, or it may be between a set of positive voltage limits or a set of negative voltage limits. Calibration curves are customarily plotted on semilog paper, with electrode potential linearly scaled and ion activity logarithmically scaled. Generally, the potential developed per order of magnitude (per decade) of ion activity change is 59.2

mV for monovalent ions, and it is 29.6 mV for divalent ions. With increased ion activity, the potential developed by the electrode becomes increasingly positive when cations are sensed, and it becomes increasingly negative when anions are sensed.

Activity is not the same as concentration. The activity of an ion is strongly influenced by the total ionic strength of the solution. Concentration, however, is usually the parameter to be determined. This is done by adding an *ionic strength adjustor* to samples and standards.

An ion-selective electrode generally looks very much like a pH electrode. It is the nature of the sensing element that gives an ion-selective electrode its characteristics. Several types of such elements are used (see Figure 3-11). A *glass membrane,* specially formulated, can be used for sodium ion concentration measurements. *Solid-state inorganic-salt-based membranes* are typically used for concentration measurements of chloride (Figure 3-11a), fluoride (Figure 3-11b), bromide and copper. *Gas-sensing* electrodes, used for measurements of ammonia, carbon dioxide and sulfur dioxide (Figure 3-11c) incorporate a gas-permeable membrane and a combination pH electrode with internal buffer solution. The gas of interest passes through the membrane and dissolves in a thin layer of buffer solution surrounding the combination pH electrode, causing the pH of the solution to change. *Polymer membrane* electrodes employ sensing membranes made from various ion-exchange materials. The chosen material is mixed with a polymer matrix and allowed to solidify. The solidified matrix is then sealed to the end of an epoxy tube. The potential developed at the surface of the membrane can be related to the concentration of the species of interest.

Ion-selective electrodes usually have bodies of glass, polymer, or epoxy, and are available mostly as combination electrodes, sometimes with separate reference electrodes. Reference electrodes can be of the single-junction type, usually Ag/AgCl with a capillary-like liquid junction or porous-ceramic junction, or they can be of the double-junction type, having an inner chamber filled with a standard solution, a porous ceramic plug between inner and outer chambers, and an outer chamber filled with the bridging solution (typically KCl). Some electrode designs are intended not to need refilling; this can be accomplished, for example, by using a gel instead of a liquid for one or both of the solutions.

The *interferences* (types of ions that, if present, or if present in significant amounts, would obscure a reading of the activity of the ion of interest) must be considered for each specific-ion electrode. They are stated in applicable manufacturers' literature. In many cases the effects of interferences can be minimized by appropriate sample preparation. Also, the temperature of the measured fluid must be known and readings corrected (or meter operation adjusted) accordingly. Specific-ion ("pIon") meters are standard-

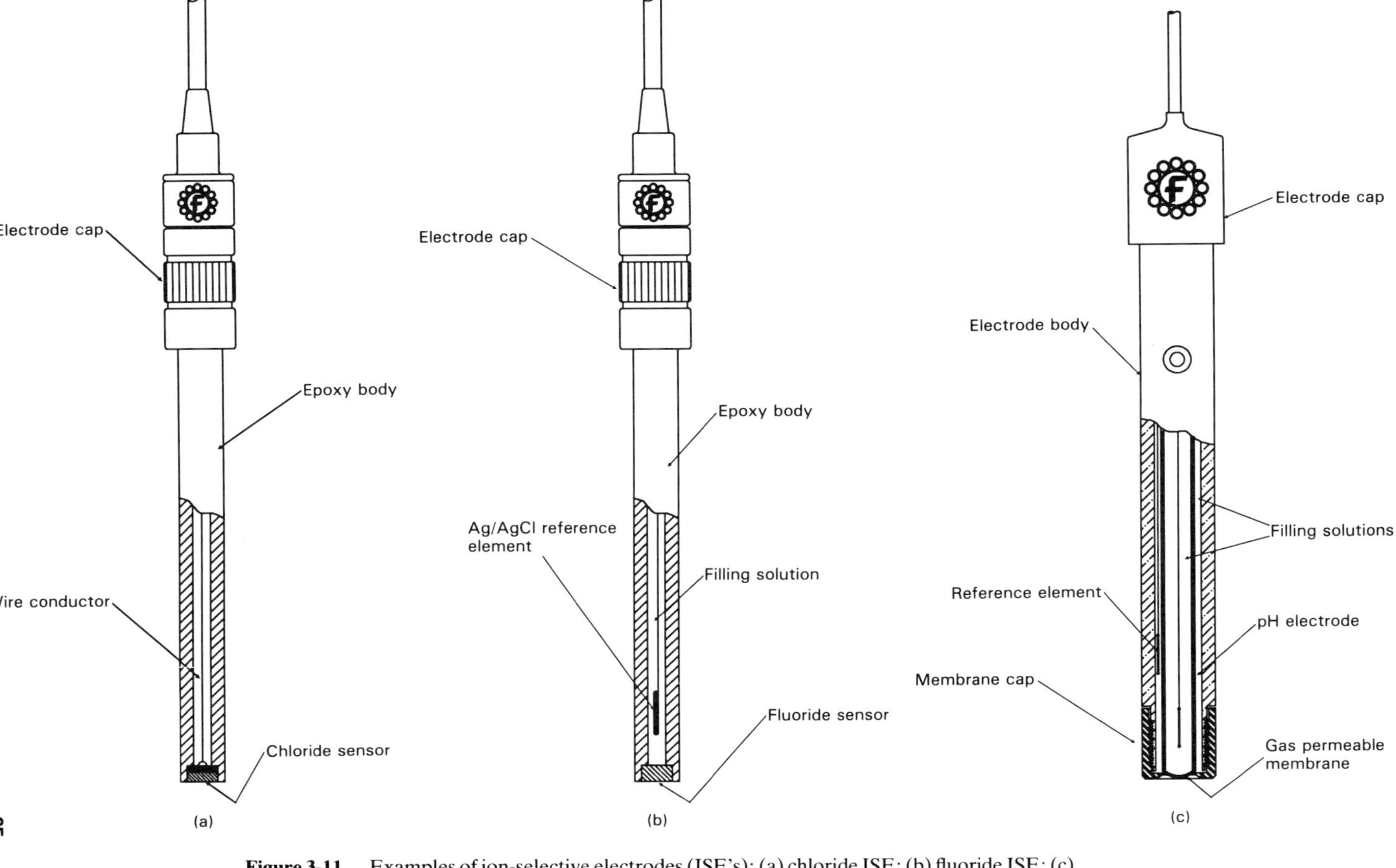

Figure 3-11. Examples of ion-selective electrodes (ISE's): (a) chloride ISE; (b) fluoride ISE; (c) gas sensing ISE (ammonia, carbon dioxide and sulfur dioxide). (Courtesy of Fisher Scientific.)

ized by immersing the electrode in a solution of known activity of the ion of interest.

Sample-preparation methods specified for the use of each type of ion-selective electrode have been developed, are stated in applicable literature, and must be followed. In addition to the ion-strength adjustor, explained above, a pH adjustor may sometimes have to be added to the sample to avoid damage to the electrode from highly acidic or highly alkaline samples.

Developmental efforts have recently been directed at modifying field-effect transistors (FET) into ion-selective sensors by chemical grafting of the silica insulator. By such grafting, an ion-selective FET (ISFET) as well as a reference FET (REFET) have been constructed in several laboratories and tested, with promising results.

3.5 COULOMETRIC INSTRUMENTS

In *coulometric* instruments, the quantity of electricity required to carry out a chemical reaction is measured. This quantity is based on *Faraday's law,* by which the reaction of one gram-equivalent weight of a substance will be effected by the passage of a quantity of electricity of 96 496 coulombs; this assumes that the reaction is 100% current efficient. The latter assumption as well as the requirement that only one overall reaction occurs are fundamental to coulometry.

Two principal methods are used in coulometry. The *constant-current* method requires a closely controlled-current source as a power supply, which is connected across two *generator electrodes.* The latter are placed in the cell assembly, which also contains the working electrode and the reference electrode and a stirrer, which is normally driven by magnetic coupling to a permanent-magnet stirrer located outside the cell. The sample solution is placed within the cell. The current from the constant-current source is set to a value that permits the electrolysis to be completed within some reasonable period of time (3 min or less). A timer is started when the power is turned on to the generating electrodes that effect the electrolysis of the sample. The timer is stopped when the electrolysis is completed, as indicated by the potential from the working (and reference) electrode. The number of coulombs required for the electrolysis is then computed from knowledge of the current and of the elapsed time. An excess of a redox buffer substance is usually added to the sample to keep the potential developed from getting high enough to cause an unwanted reaction; the buffer will also act as an intermediate in the reaction. The area of the generating electrodes must be large enough to establish a current density low enough to keep electrode polarization within limits required for 100% current efficiency. A simplified instrument schematic is shown in Figure 3-12.

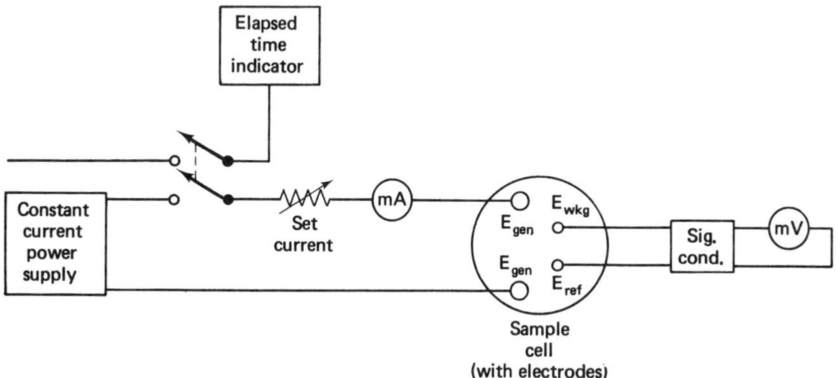

Figure 3-12. Schematic of constant-current coulometer.

The *controlled-potential* method requires continuous adjustment of the power supply current to keep the working-electrode potential constant as compared to the reference electrode. The current is monitored and the integral of the current over a selected time period is displayed together with the current itself and the control potential. A typical controlled-potential coulometer uses a multiple platinum-gauze cathode (to provide a large electrode area), a platinum-wire anode, and a calomel reference electrode. Current integration is performed electronically. A typical block diagram is shown in Figure 3-13.

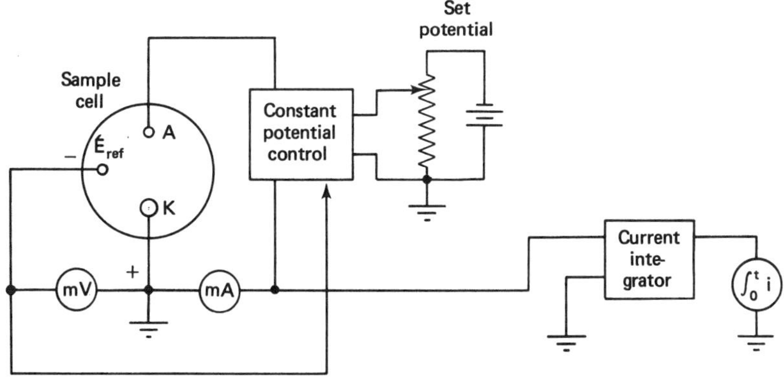

Figure 3-13. Schematic block diagram of controlled-potential coulometer.

3.6 POLAROGRAPHS

An effect known as *concentration polarization* occurs in most electrochemical cells. A region in the solution in contact with the electrodes acts as a depletion layer where the ions of the solution are discharging at the electrode, and as an excess layer when the electrode is ionizing. The current at the electrode is then not only dependent on ion activity and electrode potential; it will include a component due to the rate of diffusion of ions from the main portion of the solution into that layer (*diffusion current*). For many electrometric methods the concentration polarization and diffusion current are minimized by making electrode surfaces relatively large. In a polarograph the concentration polarization is maximized by using an electrode of extremely small surface area (*microelectrode*) so that the current reading is due to the diffusion effect to the greatest extent possible.

The classical polarographic microelectrode is the *dropping mercury electrode,* used as the cathode in the polarographic cell in conjunction with a mercury-pool electrode, covering the bottom of the cell, that is used as the anode. The dropping mercury electrode is, essentially, a glass capillary at the bottom of, and fed by, a mercury reservoir. The drop of mercury that grows at the bottom of the capillary and finally separates from it and falls into the mercury pool acts as the cathode surface. It has the advantage of being very small in surface area as well as being replaced constantly so that unwanted deposits have no time to form on its surface and create an unwanted polarization effect.

The operation of the polarograph, in its simplest form, involves applying a voltage, which varies in amplitude as a function of time in a predetermined manner, across the electrodes and measuring the current flowing through the circuit (see Figure 3-14). The current will typically increase with increasing potential but will reach a plateau due to the diffusion current of a particular

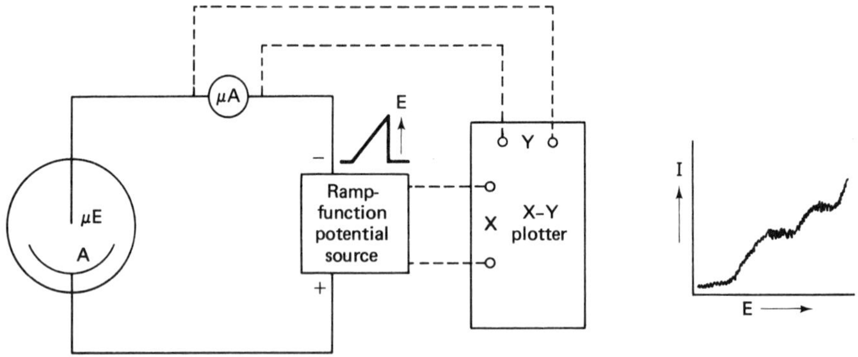

Figure 3-14. Polarograph—schematic block diagram (with example of a polarogram).

ion. When more than one ion is present in a solution, successive plateaus at different voltages and currents will be observed.

In many applications, the mercury-pool anode is replaced by a calomel reference electrode whose cell is connected to the measuring cell by a salt bridge, giving the two-cell/agar salt-bridge tube assembly the shape of an H (H-cell). The microelectrode may also be other than of the dropping mercury type (e.g., gold, platinum, carbon, or carbide). When the current–voltage curve is differentiated over a potential scan, a polarogram is obtained that shows peaks (*derivative polarography*) where the normal *polarogram* would show plateaus. Since the peaks obtained by derivative polarography are usually more pronounced than the plateaus, this method is quite attractive. Some polarographs employ a microelectrode and a counterelectrode (e.g., mercury pool) as well as a reference electrode and can then use the signal from the reference electrode to control the applied scan voltage very precisely.

A modern polarographic sensor, intended primarily for monitoring oxygen in process streams, is shown in Figure 3-15. A special material is stretched over the sensing membrane to minimize a problem common to such sensors: electrolyte consumption. The material does not inhibit oxygen diffusion. The sensor is disposable, after a useful life of six to eight months. A thermistor is incorporated in the sensor for temperature compensation purposes. A special activation cycle is provided by the associated electronics. In this cycle, a voltage pulse of positive polarity is periodically applied to the cathode. This tends to reduce drift and nonlinearity and improves the response time.

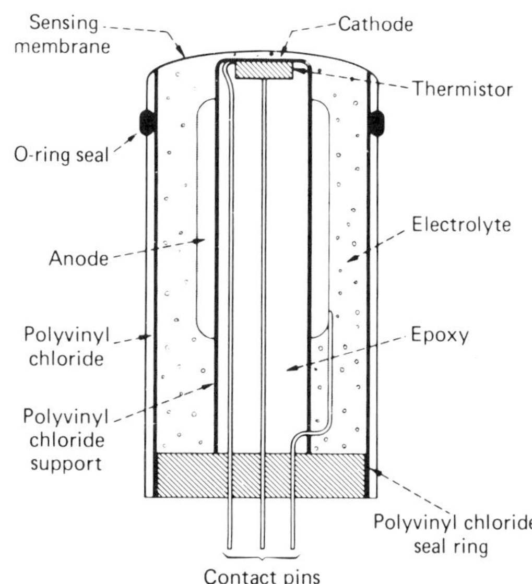

Figure 3-15. Polarographic sensor that responds to partial pressure of oxygen. (Courtesy of Teledyne Analytical Instruments.)

3.7 ELECTROCHEMICAL GAS SENSORS

A variety of electrochemical sensors (electrochemical cells) have been developed for gas analysis. They provide a direct measurement of the concentration of a gas in a gas mixture, usually either in ambient air or in a process stream (the latter includes flue emissions). Some sensors were designed specifically for measurements of a single gas, such as oxygen. Other designs can measure two gases simultaneously, and some generic designs can be modified by their manufacturers to measure one of several gases. A few examples are described in this section.

The *micro-fuel cell* (Figure 3-16) is related to fuel cells used as power generators (e.g., on the Shuttle orbiter). It is self-generating (does not need a power supply), but does not operate until a fuel (oxygen) is supplied to it. Its operation is based on the reduction of oxygen at the sensing electrode, resulting in a current that is directly proportional to the partial pressure of oxygen in the sample gas. This current is dependent on the diffusion path. The rate of diffusion of oxygen to the sensing electrode is controlled by movement through the sensing membrane and the thin film of electrolyte between the membrane and the electrode. The sensor is temperature-sensitive; however, good temperature compensation is provided by use of a negative-temperature-coefficient thermistor. The sensor operation is not affected by vibration or attitude (position) and has a measuring range of 1 ppm to 10 atmospheres of oxygen. The associated electronics and read-out unit is available in various versions, including recessed wall-mounted, and portable, battery-operated ones. The fuel-cell principle is also used in some sensors for gases other than oxygen.

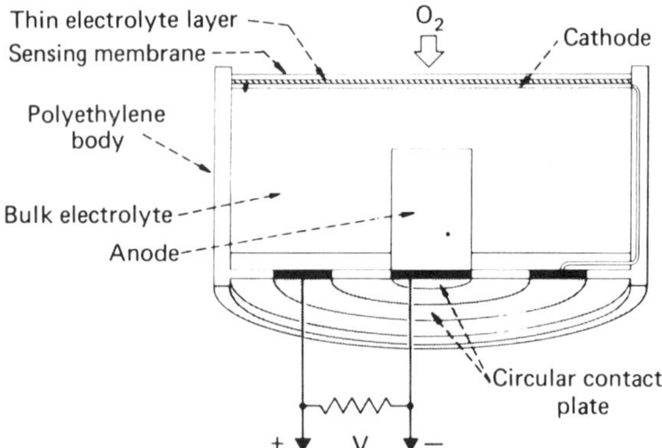

Figure 3-16. An electrochemical sensor for oxygen analysis: the micro-fuel cell. (Courtesy of Teledyne Analytical Instruments.)

Among the most critical measurements in flue gas emissions and combustion systems are those of nitrogen oxides (NOx) and sulfur dioxide (SO_2). Figure 3-17 shows an electrochemical sensor, with associated units, for the simultaneous monitoring of both gases. The sensor is a metallized membrane electrochemical cell. A liquid electrolyte is separated from the sample gas by a permeable membrane which has a thin layer of metal sputtered onto its electrolyte side. The metallized layer, which acts as an electrocatalytic surface, forms the sensing electrode. It is held at a fixed voltage potential with respect to a reference electrode. A third, auxiliary, electrode prevents polarization of the sensing electrode. The three electrodes are connected by a potentiostat. The output current is a function of the partial pressure of a particular gas that permeates through the membrane. It has been shown that a voltage of 0.5 to 1.0 V between sensing and reference electrodes makes the sensor respond only to SO_2 and NOx. In order to produce separate current outputs for these two gases, the sensing electrode is split into two sections that are electrically insulated from each other, and each section is connected in a separate circuit with its own reference and auxiliary electrode. Different potentials are used in each portion of this dual electrochemical cell, so that two independent current outputs are produced, one for each gas. Two equations with two unknowns (concentration of SO_2 and concentration of NOx) are solved simultaneously

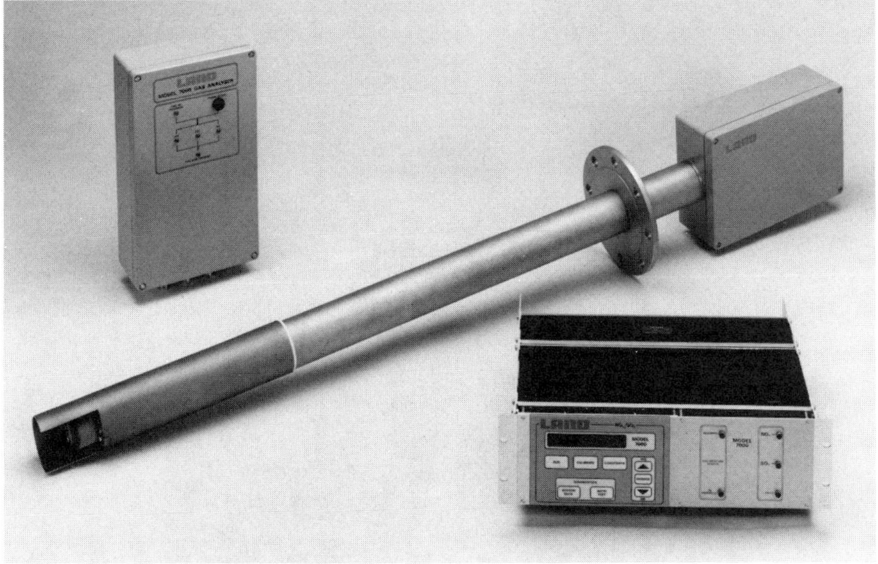

Figure 3-17. Electrochemical NOx/SO_2 sensor with associated calibration and electronics/read-out. (Courtesy of Land Combustion.)

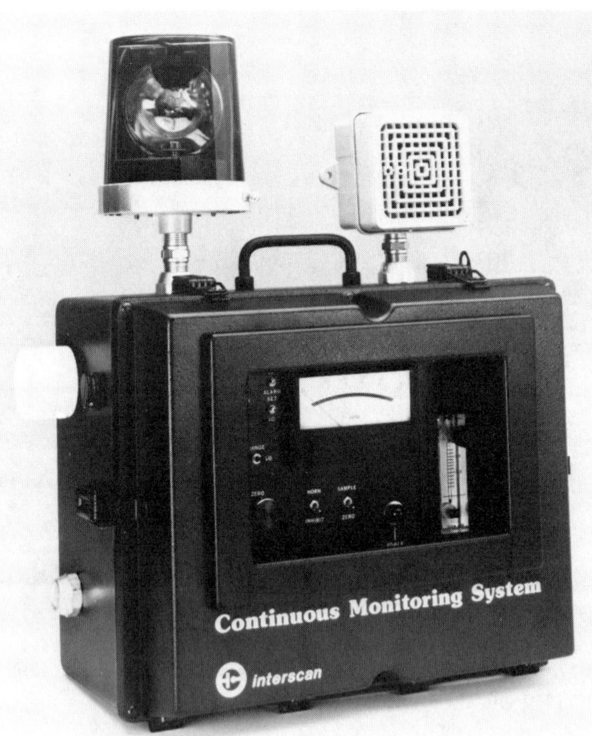

Figure 3-18. Toxic gas analysis system: (a) electrochemical sensor; (b) compact analysis system with alarm devices. (Courtesy of Interscan Corporation.)

for the calculation of the two concentrations. A sintered stainless-steel filter at the probe tip prevents the accumulation of particulates in the sensor. Microprocessor-based electronics, used to produce the two output signals, are contained in a housing at the other end of the probe. Cabling and another microprocessor in the electronics and read-out unit, typically installed in the control room, establish the rest of two current loops.

A continuous monitoring system for toxic gases, as well as the sensor used in it, are illustrated in Figure 3-18. Depending on user requirements, the sensor and system can be factory-adjusted to respond to CO, NO_2, NO, H_2S, SO_2 or Cl_2. The sensor (U.S. Patent No. 4 017 373) is an electrochemical cell operating under diffusion-controlled conditions. The bias voltage applied between the sensing and nonpolarizable reference electrodes, as well as the choice of electrolyte (in an immobilized matrix) are the primary factors in making the sensor specific to one of the gases. The diffusion-limited output current, which is directly proportional to gas concentration, is converted into a voltage for meter or recorder display. The manufacturer supplies information about interfering gas for each type of sensor, and about the severity of such interferences. Severely interfering gases have to be scrubbed out by selective chemical filters.

4

SOLID-STATE GAS ANALYZERS

Gas analysis (the measurement of the concentration of one gas in a mixture) is one of the primary objectives of analysis instrumentation, especially in industrial applications. The term "solid state" has become a buzzword, and there is a thin line that separates other instruments (such as some types of electrochemical cells) from gas sensors that truly can be termed solid state. This chapter covers the few types of sensors, commonly accepted as solid state, that are commercially available, as well as some sensors currently in a more or less promising stage of development.

The first commercially available solid-state gas analyzer (successfully produced for many years) is the *zirconia oxygen analyzer* (zirconia is a short term for zirconium dioxide). Basically, this oxygen sensor consists of a tube of zirconium dioxide (in some designs stabilized with another element such as calcium), closed at one end. Electrodes of porous platinum are coated onto the inside as well as the outside of the tube. The tube is heated electrically to a high temperature and the temperature is monitored by a platinum resistance thermometer which also provides a signal to a proportional temperature controller. The measured gas is applied to the outside of the tube and a reference gas mixture (containing a known amount of oxygen) is applied to the inside of the tube; alternatively, the measured gas is applied inside the tube and the reference gas outside the tube, as in the sensor shown in Figure 4-1. At high temperatures (600 to 900 °C) the zirconia tube becomes an electrolytic conductor due to the mobility of the oxygen ions. The tube surface toward the higher oxygen partial pressure acts as the anode, whereas the surface toward the lower oxygen partial pressure acts as the cathode. At high temperatures, the oxygen molecules at the anode absorb electrons, while the reverse process takes place at the cathode. The voltage produced by the sensor is (under open-circuit conditions) given by the Nernst equation:

$$E = KT \log p_{ref}/p_{meas}$$

where K = a constant
T = absolute temperature
p_{ref} = partial pressure of oxygen in the reference gas
p_{meas} = partial pressure of oxygen in the measured gas

The reference gas can be a gas mixture containing a closely controlled percentage of oxygen. Adequate measurements can also be obtained by using ambient air as reference, since air contains about 210,000 ppm of oxygen. In some simpler designs, the inside of the tube is vented to the ambient atmosphere and the measured gas is applied to the outside of the tube. The output signal is fed to a closely-coupled preamplifier with a very high input impedance. The amplified signal is then further amplified or otherwise conditioned for the desired type of display.

Extensive research has been carried on, in many countries, to develop solid-state gas sensors of various types. These include primarily semiconductor sensors, employing thick or thin films of metal oxides. Additionally, field-effect transistors (FETs) have been modified by chemically altering their gates

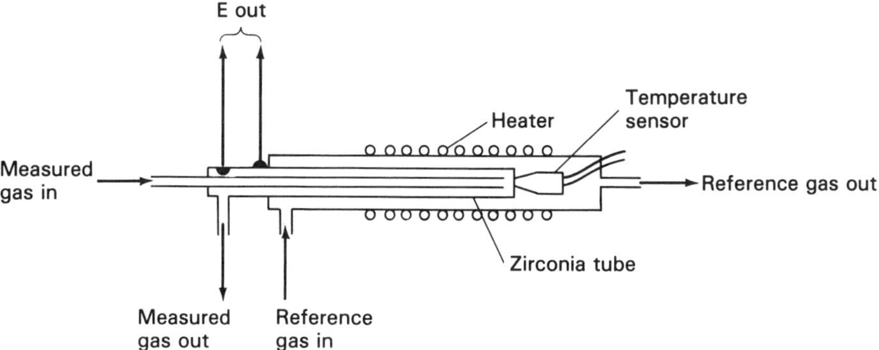

Figure 4-1. Example of heated-zirconia oxygen sensor.

to respond to specific gases (GasFETs). Proton-conducting polymer films are used for the detection of a few gases, mainly hydrogen. Some types employ a reference electrode. Many types require temperature control (based on a built-in temperature sensor) with or without the use of a heater.

A few designs have reached the production stage; others are in the pilot-production stage; and many more admittedly (by their developers) need more work, for example, to increase their sensitivity, to make them more specific to one gas; and to make them producible. A few examples are described below.

The *titanium-oxide sensor* (titania resistor) is usable as an oxygen analyzer. The titania film changes resistance when oxygen molecules diffuse into the material. Since the sensor can also act as a thermistor, its temperature is controlled within close limits.

The *tin dioxide gas sensor* (SnO_2 sensor) is a semiconductor sensor that can respond to various gases. A typical design contains an SnO_2 film on a high-temperature insulating substrate (e.g., sapphire), bonded to a ceramic heating element. Platinum electrodes are applied to the film. To make the sensor (relatively) specific to one gas, the film is covered with a semipermeable (permselective) membrane or a filter, or the SnO_2 is combined with additives. SnO_2 sensors have been built for the detection of CH_4 in domestic gases, for measuring concentrations of NOx in air, for H_2S and for CO, hydrogen and oxygen measurements (relatively more development work is reported for CO sensors).

Tungsten trioxide (WO_3) based sensors, especially metal-oxide semiconductor (MOS) sensors, have been used successfully as H_2S detectors. Also, a platinum-activated tungsten oxide sensor was developed to measure low concentrations of hydrogen in air.

Proton-conducting polymers have also been studied for use in gas analyzers. One commercially available design uses a polymer film, both sides of which are coated with platinum electrodes, as a hydrogen sensor. The membrane separates the reference hydrogen gas (with known concentration) from the measured gas, and protons are conducted from the side with higher partial pressure to the side with lower partial pressure. The resulting charge imbalance across the membrane is converted into the sensor output voltage.

5

THERMAL ANALYSIS INSTRUMENTS

This section discusses the principles of three classical methods of thermal analysis (differential thermal analysis, differential scanning calorimetry, and thermogravimetry) as well as two additional types of sensors employing thermal effects but not usually considered as thermal analyzers (heat-of-combustion sensors and thermal conductivity cells).

Most of the classical thermal analysis methods are aimed at displaying heat effects on a substance, associated with chemical and physical changes that the substance undergoes when heated. These effects may be *endothermal* (characterized by the absorption of heat) or *exothermal* (characterized by the development of heat). Causes of such effects include vaporization, boiling, sublimation, crystalline structure transformations, fusion, and such chemical reactions as oxidation, reduction, dehydration, decomposition, and direct combination. The purity of a substance can often be determined by measuring its boiling point or freezing point, usually with reference to the boiling or freezing point of the "standard" of that substance. Some of the changes that substances undergo with heating (at a known rate with reference to time) can also result in changes of sample mass. Finally, some thermally caused changes are detected by controlled cooling instead of heating.

5.1 DTA, DSC, AND TGA

In the *differential thermal analyzer* (*DTA*), the temperature of a furnace is increased or decreased linearly with time, using a temperature sensor such as a thermocouple as input to a temperature programmer/controller in such a manner that a preselected heating rate (e.g., 20 °C per minute) or cooling rate is maintained exactly. The furnace contains a cell assembly with two cavities, one for the sample material, the other for a reference material against which the thermal behavior of the sample is to be compared. The sample temperature and the reference-material temperature are measured by a differentially connected pair of temperature sensors, typically a differential thermocouple. When the thermal behavior of both materials is identical, the output of the differential thermocouple will be zero. When the sample exhibits an endothermic or exothermic event not experienced by the reference material, the differential thermocouple will provide an output whose magnitude is indicative of the heat generated or absorbed, and whose polarity indicates whether the event was endothermic or exothermic. A schematic representation of a DTA is shown in Figure 5-1. The (amplified) outputs of the differential thermocouple and of the heater-temperature monitoring sensor are usually displayed, the former vs. the latter, on an X-Y plotter. DTA equipment is capable of heating as well as cooling at a controlled linear rate.

The *differential scanning calorimeter* (*DSC*) is similar to the DTA except that temperature differences between sample and reference are determined in

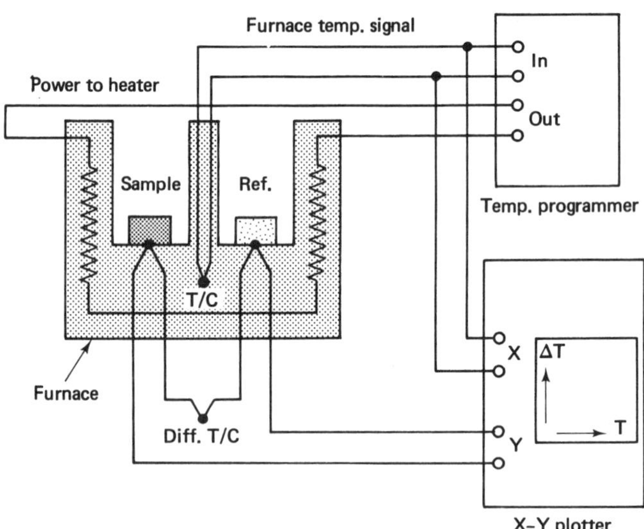

Figure 5-1. Differential thermal analyzer—simplified schematic diagram.

a close-loop mode in the DSC, whereas they are detected in an open-loop mode in a DTA. In the DSC the sample and reference materials are heated by individual heaters; both heaters are initially programmed and controlled to heat or cool the materials at the identical rate. Temperature sensors are provided to measure both sample and reference temperatures. When the sample temperature differs from that of the reference material, due to an endothermic or exothermic event, the heat seen by the sample is rebalanced until its temperature matches that of the reference. The change in heater current required to effect this rebalance is then a measure of the heat developed or absorbed by the sample.

A *thermogravimetric analyzer* (*TGA*) measures the weight changes that a sample undergoes, due to thermal effects, while being heated or cooled at a known rate. The mass of the sample, which is inserted in a furnace, is measured continuously by a sensitive balance that provides an output signal indicative of sample mass. The output of this *thermobalance* is then recorded vs. a signal representative of sample temperature.

5.2 THERMAL CONDUCTIVITY CELLS

The *thermal conductivity cell* (thermal conductivity detector) is used for the analysis of gases on the basis of the thermal conductivity of a sample gas with reference to the thermal conductivity of a reference gas. The device (see Figure 5-2) consists of four identical helical filaments (typically gold-sheated

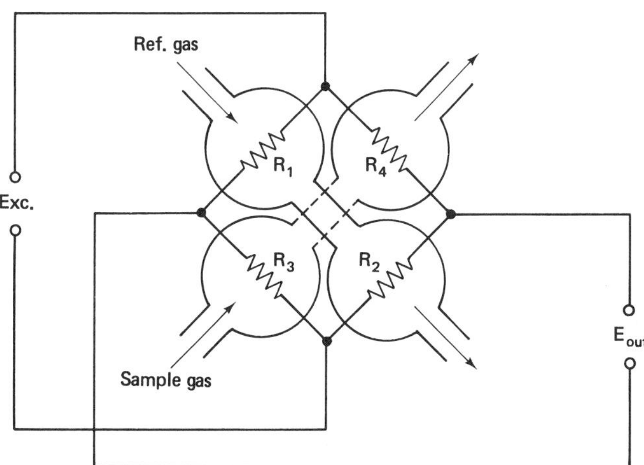

Figure 5-2. Thermal conductivity cell—schematic diagram.

or Teflon-covered tungsten wire) connected as four arms of a Wheatstone bridge in the manner shown. Each filament is positioned at the center of a small cavity. The two sample-gas cavities are connected by a passage, but they are isolated from the reference-gas cavities and their connecting passage. The excitation power furnished to the bridge is large enough to cause heating of the four filaments. The temperature of the filaments and, hence, their resistance, is a function of heat loss into the gas. The bridge is adjusted so that it is balanced, and the bridge output voltage is zero when the thermal conductivities of both gases are the same. When the composition of the sample gas changes, its thermal conductivity changes, and the resistance of filaments R_3 and R_4 changes accordingly, whereas the resistance of filaments R_1 and R_2 remains constant (provided that other factors, especially temperature, pressure, and flow rate of both gases, are the same).

Thermal conductivity cells are used for gas analysis in gas analyzers as such and as detectors in many models of gas chromatographs (see Section 8.1). They are used primarily in applications where one gas is to be analyzed in a stream containing one other gas. Some examples are CO_2, SO_2, or NH_3 in either N_2 or air; H_2 in Ar, N_2, Cl_2 or HCl; or Ar in O_2. Figure 5-3 shows examples of gas-relative thermal conductivities. When used in a gas chromatograph, the carrier gas acts as reference gas and the effluent from the column is the gas to be analyzed.

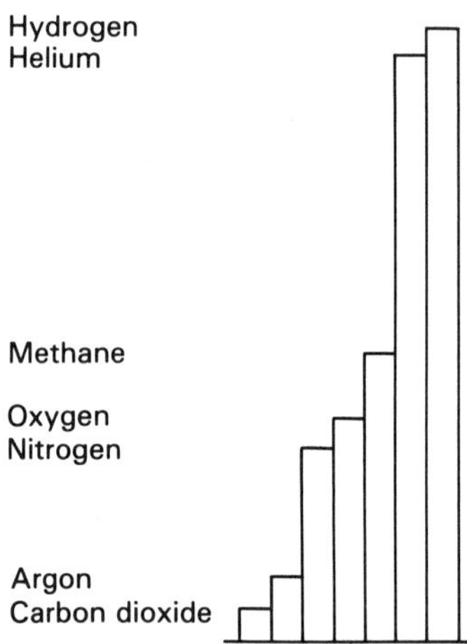

Figure 5-3. Examples of relative thermal conductivity of gases. (Courtesy of Teledyne Analytical Instruments.)

Since the thermal conductivity cell is, electrically, a Wheatstone bridge, potentiometer-type controls typical for such bridges are always incorporated in gas analyzers using such cells; they control zero as well as span. The instrument output, usually isolated, can be a dc voltage or current. Figure 5-4 shows a typical gas analyzer using a thermal conductivity cell.

Figure 5-4. Thermal-conductivity gas analyzer. (Courtesy of Teledyne Analytical Instruments.)

5.3 HEAT-OF-COMBUSTION SENSORS

Heat-of-combustion sensors exist in a variety of designs. Their principal application is in the determination of combustible-gas concentration in a gas mixture. One type employs a catalyst, of special composition, which causes oxidation of carbon monoxide to carbon dioxide. The oxidation raises the temperature of the catalyst. This temperature rise is measured with reference to the temperature of the gas mixture at the inlet to the sensor, and is proportional to the concentration of the combustible gas in the mixture. Another type of sensor uses a platinum winding coated with ceramic; the ceramic surface is treated with a catalyst. The sensor also contains a reference platinum winding of the same resistance, also coated with ceramic, but without a catalyst on its surface, and not in contact with the measured gas. The two windings are connected as two arms of a Wheatstone bridge. Excitation power is high enough to cause heating of the windings. When combustible gas comes in contact with the catalyst on the sensing winding, combustion occurs and raises the temperature and, hence, the resistance of that winding. The resulting bridge unbalance voltage is proportional to the concentration of combustible gas in the gas mixture.

Platinum, itself, acts as a catalyst with certain combustible gases. This principle is used in some sensors containing two heated platinum filaments, one in contact with the measured gas mixture, the other shielded from the gas mixture and used as reference. Combustion at the heated filament raises its temperature, hence its resistance, and the output of the bridge circuit into which both filaments are connected is indicative of combustible-gas concentration.

All such sensors contain a flame arrestor of some type to prevent spreading of the combustion. The output of the sensor is usually displayed on a meter which is calibrated in terms of percent of the specific combustible gas measured. Many such meters are additionally calibrated in *percent lower explosive limit* (*%LEL*), based on the lower explosive limit established for the combustible gas intended to be measured by the instrument. The LEL is typically a few percent of the gas in a gas mixture.

6

IONIZATION ANALYSIS INSTRUMENTS

The ionization principle is employed in devices used as detectors in such instruments as gas chromatographs and total hydrocarbon analyzers. These detectors are used almost exclusively for the analysis of organic compounds in gases or (if the sample is a liquid) in vapors; solid samples can be converted into gaseous form by pyrolysis. Only a few inorganic compounds can be analyzed by some types of ionization analyzers.

6.1 FLAME IONIZATION DETECTORS

The *flame ionization detector* (*FID*) consists of a small chamber in which a hydrogen-fed flame burns in air or oxygen. The sample gas is introduced into the chamber by letting it mix with the hydrogen that feeds the flame. Hydrogen is used because it produces very few ions, and so provides a good signal-to-noise ratio. The ionic fragments and free electrons produced by the combustion of oxidizable carbon atoms produce an ionization current between two electrodes in close proximity to the flame. The (very small) current from the electrodes is then detected and amplified by an electrometer–amplifier, typically a solid-state amplifier with a junction FET front end. The ionization current is proportional to the flow rate of the sample and to the number of carbon atoms in the compound. In most designs the burner itself acts as one electrode; the other electrode is above the flame.

Figure 6-1 shows an FID; this design is used as detector in the total

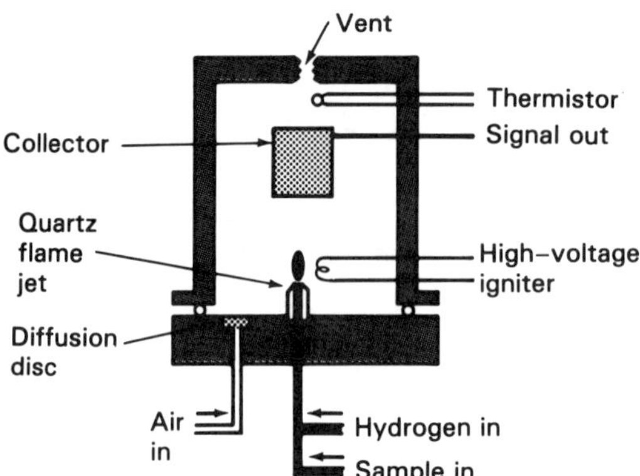

Figure 6-1. Flame ionization detector (FID). (Courtesy of Teledyne Analytical Instruments.)

hydrocarbon (*THC*) analyzer shown in Figure 6-2. The detector is contained in a stainless-steel housing. A mixture of hydrogen (40%) and nitrogen (60%) is used as fuel to be burned at the quartz flame jet. Air is added to supply the

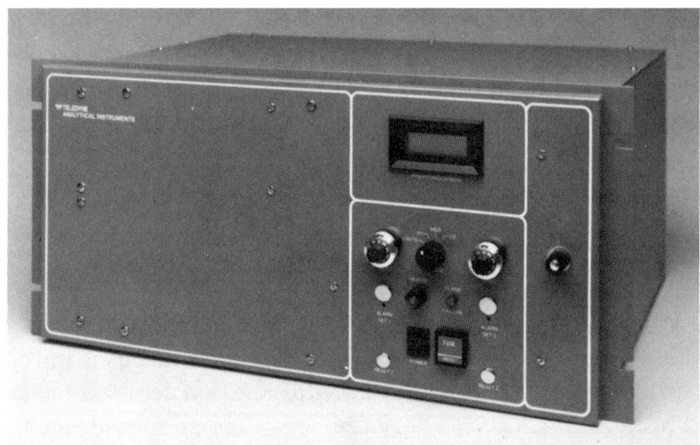

Figure 6-2. Total hydrocarbon analyzer using an FID. (Courtesy of Teledyne Analytical Instruments.)

oxygen needed for combustion, as well as to sweep away the combustion products. The continuously flowing sample is mixed with the hydrogen-nitrogen fuel. As the mixture passes the burner tip, the heat of the flame induces the cracking of any hydrocarbons present into ionic species. The platinum-wire igniter also acts as one electrode, and the basket-shaped collector acts as the other electrode; current flows through an external circuit between them when ions are present in the flame. The instantaneous current between the electrodes is proportional to the number of ions present in the flame. A thermistor acts as sensor for temperature control. The flow rates of sample, fuel, and air are critical for optimum accuracy. Flow rate control is achieved by valves, restrictors, regulators, and pressure gages in the analyzer (Figure 6-2). The analyzer provides three switch-selectable ranges (0–10, 1–100, 0–1000 ppm, or, optionally, 0–1, 0–10, and 0–100 ppm THC, methane equivalent), range and "flame-out" indicator lights, an LCD meter readout, and alarm capabilities.

6.2 PHOTOIONIZATION DETECTORS

In a *photoionization detector* (*PID*), a source of photons having relatively high energy is used to ionize the molecules of the compound being analyzed. The resulting ions are then made to cause a current flow between two electrodes. Ultraviolet photon sources have been used in some designs; other sources, such as high-frequency (RF) currents, inductively coupled, are also used, as in the example of a PID shown in Figure 6-3. The PID is often used as detector in gas chromatographs.

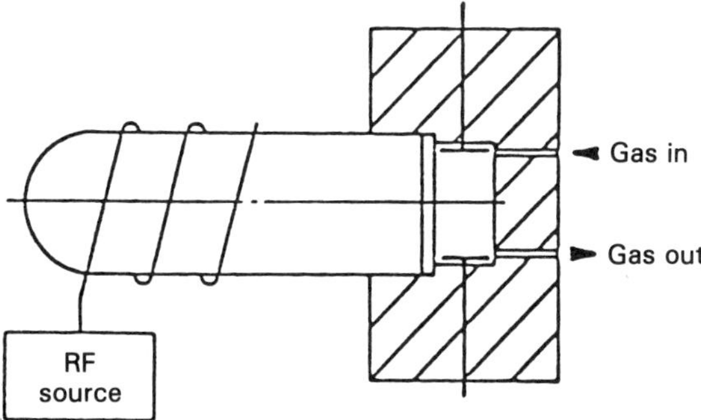

Figure 6-3. Photoionization detector. (Courtesy of Photovac, Inc.)

6.3 OTHER IONIZATION DETECTORS

Nuclear radiation from a radioactive-isotope source is used to cause ionization in some other types of detectors. One design incorporates a source of alpha particles within a cylindrical housing into which a center electrode is partially inserted, while a portion of the housing's wall acts as the other electrode. A polarizing voltage across the electrode provides for the required migration of ions, formed by collision of alpha particles with constituents of the sample gas, and the resulting ionization current. The *helium ionization detector* receives the sample gas mixed with helium carrier gas. An H^3 (tritium) foil is used as particle source as well as the anode, and a high polarizing potential is applied between it and the cathode. The field gradient and radiation due to H^3 decay raise the helium to a metastable state and the energy stored in this medium ionizes the molecules in the sample flowing through the chamber. Compounds that contain an electronegative group and thus have an affinity for free electrons can be analyzed by an *electron-capture detector.* A typical design uses nitrogen as carrier gas and a source of beta particles (e.g., H^3, Ni^{63}), which also acts as one of the two electrodes across which a polarizing voltage is applied. The beta particles ionize the nitrogen molecules and establish a field of low-energy electrons that can then be captured by the electron-capturing components of the compound being analyzed. The removal of an electron from the medium between the electrodes causes a reduction of the baseline ionization current, and this current change then characterizes the compound. A design version of an electron-capture detector (*ECD*) has been reported using charged parallel electrodes instead of a radioisotope source to provide the required electron field density.

7

PHOTOMETRIC ANALYSIS INSTRUMENTS

7.1 PHOTOMETRIC SENSING METHODS

A photometer is (by literal translation from the Greek) a light measuring device. Photometric analyzers do invariably employ a light sensor (photodetector, see Section 9.2.4); however, in most applications the analysis is based on the *interaction of light with substances* and, in some applications, on the emission of light from substances. Figure 7-1 illustrates the basic sensing methods.

A number of important material-characterization methods are based on the interaction of light with substances. These include transmittance, reflectance, 90° light scattering (nephelos), fluorescence, and refraction of light emanating from a light source, usually in the form of a collimated beam. These methods are used mainly in detecting the content and concentration of particulates in liquids and gases; however, some other physical and chemical properties can also be determined by such methods. Additionally, there are methods that use light emitted from a sample that has been manipulated in some manner, such as mixing it with certain reagents or passing it through a flame.

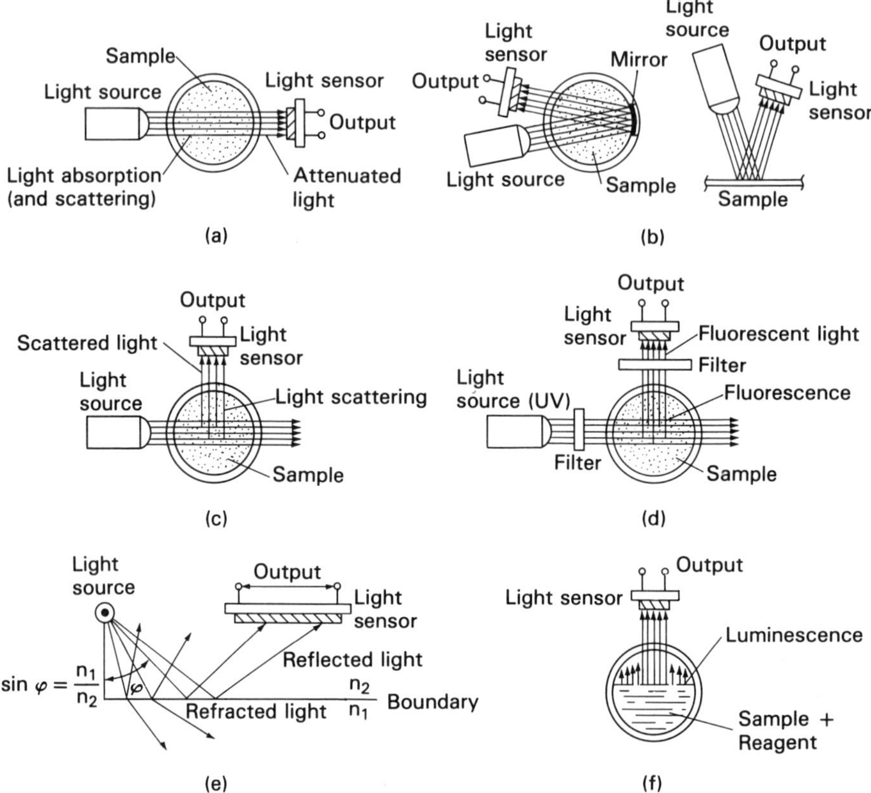

Figure 7-1. Photometric sensing methods: (a) transmittance; (b) reflectance; (c) nephelos; (d) fluorescence; (e) refraction; (f) light emission.

Major analytical applications of these methods, described in more detail in subsequent sections of this chapter, include the following: both transmittance and reflectance are used for color determinations and for related compositional analysis by *spectrophotometry* (see Section 9.1); transmittance is used for opacity monitoring, for densitometry (this term refers to *optical density*) as well as for turbidimetry (transmission turbidimetry); nephelos is used in turbidimetry (nephelometric turbidimetry) and in nephelometry (a form of turbidimetry but applied mainly to air); fluorescence is used in fluorimetry, and refraction in refractometry. Light emission methods are used in chemiluminescence analyzers and in flame photometers.

7.2 COLORIMETERS

Two different methods are used to measure color: the *direct method* and the *comparison method*. Both methods are employed to measure the color characteristics of either reflected light (*reflectance mode*) or transmitted light (*transmittance mode*). In the indirect method, the sample of light, whose color characteristics are to be determined, is collected by suitable optics, passed through either a continuously adjustable or discrete-step adjustable monochromator, and is then converted into electrical output signals by one or more light sensors. In the direct method, color is determined by means of the visual equivalence of the sample to a synthesized stimulus. Both methods require a source of illumination of known characteristics.

Two categories of instruments are used for color measurement: spectrophotometers (spectrocolorimeters) and colorimeters. *Spectrophotometers* (see Section 9.2) measure the spectral distribution of radiant energy (luminance) over the visible spectrum of light. Two basic types of this instrument are in use: (1) the *full* (or *dispersive*) *spectrophotometer* employs a continuously adjustable monochromator, such as a variable interference filter; *dispersion* refers to the (rotary) dispersion of polarized light (e.g., by varying the adjustments of Nicol prisms); and (2) the *abridged spectrophotometer,* or abridged narrow-band filter-type spectrophotometer, uses a number of separate filters, typically also interference filters, to divide the spectrum into discrete bands of wavelengths; typical filter intervals range between 20 nm (16 filters, 380 to 700 nm) and 10 nm (37 filters, 380 to 750 nm). The spectrophotometer output signals are visually displayed, most frequently graphically recorded (*recording spectrophotometer*), and can be processed by a minicomputer to provide corrections for zero and 100% line errors as well as calculations of color coordinates. Spectrophotometric measurements are usually referenced to "100% white," using a very white material such as barium sulfate to provide the reference color. To allow for fluorescence in the measured sample, however, spectrophotometers are often designed to provide readings up to "200% white." Spectrophotometers are widely used for chemical analysis (see Section 9.2.5).

Colorimeters are light source–filter–light sensor combinations which simulate the tristimulus functions of visual color perception (*tristimulus colorimeters*). Three filters and either one or three light sensors are employed to provide three separate outputs for X, Y, and Z. In their simplest form, one output is obtained for the amount of red, one for the amount of green, and one for the amount of blue in the sample. In the so-called three-color method of colorimetry, the sample is matched with a mixture of variable amounts of three components of light with different (but known) chromaticities. Three broad-band filters are incorporated in colorimeters for spectral separation. Colorimeters are used mainly for comparisons between similarly colored specimens; hence, most designs also provide outputs in terms of color difference. To further simulate human visual response the outputs are often made proportional to the cube root of luminance.

Spectrophotometric color characterization is generally more accurate than colorimetric determinations; however, interpretation of colorimetric data is considerably simpler than that of spectrophotometric data. Important performance characteristics for spectrophotometers include the spectral resolution by either the design of the continuously variable monochromator or by the number of separate filters used and their band-pass characteristics; photometric linearity is another important characteristic, together with photometric resolution and the time required to scan the complete spectrum (*scan speed*). Scan speeds of between 30 and 45 s are common, as are spectral resolution between ± 0.2 and 0.5 nm and photometric resolution to 0.01%. Photometric resolution below about 5% reflectance is generally poorer than that at higher reflectance values. It should be noted that instruments designed for transmittance cannot readily be used for operation in the reflectance mode; however, many instruments are designed to operate equally well in both modes.

For colorimeters, the linearities and sensitivities inherent in the various readouts are essential performance characteristics. Lighting and viewing geometries must be selected to fit the application; the viewing geometry is most commonly zero degrees; lighting geometries include 45°, 360° (circumferential), and spherical chambers for diffuse hemispherical viewing. In the transmittance mode, the light seen by the light sensor should include scattered as well as transmitted light to provide correct color characterization. The spectral quality of the light source (or other illumination) must be accurately known for all color measurements, and calibration stability as well as ease of calibration are important characteristics of all color measurement devices.

Colorimetric techniques are not only applied to color determinations as such; they are also used for chemical analyses of solution and gas concentrations. To measure the concentration of a specific constituent in a solution, a predetermined quantity of a distinct reagent is added to it. The solution then changes color and this color is compared with a reference color, usually by comparing it with the color of a standard sample of known concentration. Such instruments are often referred to as *color comparators*. A direct reading, in terms of mg/L or ppm can be provided by employing interchangeable meter scales. Color comparators operate either in the transmittance or reflectance mode.

Gas concentrations can be determined by impregnating a paper-like tape with a dry reagent of a type specific to the gas to be analyzed. When exposed to that gas, the spot of reagent traps a portion of the gas and changes color (the higher the concentration of gas, the darker the stain). This color is measured by a colorimeter of special design (see Figure 7-2) and compared to a standard response curve. The manufacturer of the instrument whose operating principle is illustrated in this figure manufactures such tapes in the form of interchangeable cassettes to which he applies the name "Chemcassette."

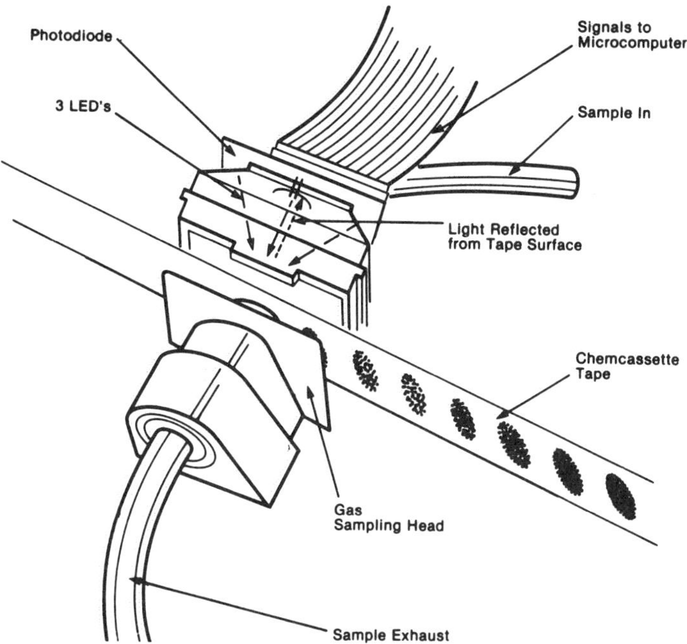

Figure 7-2. Colorimetry applied to dry-reagent gas analysis. (Courtesy of MDA Scientific, Inc.)

7.3 OPACITY MONITORS

Opacity sensing devices (sometimes called transmissometers) employ the sensing method shown in Figure 7-1a. They measure the attenuation of light in the sample fluid due to absorption and scattering (% opacity = 100% − % transmittance). The light incident upon the light sensor decreases with increasing opacity in the sample.

One of the most common applications of opacity monitors is as smoke density monitors in exhaust ducts and smoke stacks. In a typical smoke density monitor installation, the light source and photodetector are installed at precisely aligned openings in the stack or duct, 180° apart. Contamination of both optics is prevented by a continuous flow of purge gas (filtered air). An alternative is to place both the light source and sensor at one opening and install a suitable reflector at the other, 180° apart; this increases the transmittance path. The reflector assembly is also kept clean by purge gas.

Readout devices display the output of opacity monitors in various terms. Included in these is % transmittance, % opacity, and optical density. The latter is equal to the $\log_{10}$ of $1/(1 - \text{opacity})$. Percent opacity can also be shown in terms of the Ringelman scale, four rasters representing four discrete "shades of gray" in steps of 20% opacity (2 = 20% . . . , 4 = 80%) printed at 90° spacing around a hole, of the same dimension as each of the rasters, through which smoke venting from a stack is viewed (with the sun behind the observer) and its opacity compared with the opacity represented by the rasters. Opacity monitors can be calibrated by inserting standard neutral density filters into the transmission path.

Lasers have found increasing use as light sources in stack monitors as well as in other opacity-based air-quality monitors in which the light sensor is placed a sufficient distance away to allow an appropriate volume of air to be sampled (or packaged together with the laser light source for nephelometric determinations). Lasers lend themselves to such applications because they provide a well-collimated beam of linearly polarized light at a single wavelength.

7.4 TURBIDIMETERS AND NEPHELOMETERS

Turbidimeters and nephelometers are used to measure the amount of haziness in a liquid or gas, caused by suspended particulate matter. A higher turbidity value indicates more haziness, a lower value less haziness (more clarity). Higher values of turbidity are measured by turbidimeters; lower values are measured by nephelometric turbidimeters or nephelometers. Generally, turbidimeters base their operating principle on transmittance (Figure 7-1a) as it is affected by absorption and scattering of light; in some cases, only scattering is used. Nephelometers and nephelometric turbidimeters base their operating principle on nephelos, the scattering of light at 90° to the light beam (Figure 7-1c).

In a *turbidimeter* the light beam passes from the light source, through the sample, to a light sensor. If the fluid is perfectly clear, the light sensor will detect the maximum light intensity. When the sample is not perfectly clear, especially when it contains solid particles, the light beam will be attenuated while passing through the sample and the light intensity seen by the sensor will be reduced. The two main causes for this attenuation are *absorption* and *scattering* of light. Absorption occurs in accordance with *Beer's law,* which states that the absorption of light by a solution changes exponentially with the concentration (all else remaining the same). When the attenuation of the light beam is solely due to absorption, this sensing method is known as *absorptometry.* Additional attenuation will occur when the particles in the sample fluid are large enough to scatter light. The ratio of the intensities of scattered to incident light, at a specified distance, is known as the *Rayleigh ratio.* The scattered energy is maximum when the radius of the (assumed to be spherical) particle equals the wavelength of the light beam. When this radius is much larger than the wavelength, the scattered energy is nearly independent of wavelength. When this radius is much smaller than the wavelength (*Rayleigh scattering*), the scattered energy decreases with the inverse of the fourth power of wavelength. Another characteristic of Rayleigh scattering is that the intensity of scattered light, as scattered in any direction of angle θ with the incident-light direction, will be directly proportional to $1 + \cos^2 \theta$. A further characteristic of Rayleigh scattering is that the light will also be plane-polarized. The amount of polarization (of a light beam through a colloidal solution of high dispersity) depends on the size of the particles; polarization is complete when the size of the particles is much smaller than the wavelength (*Tyndall effect*).

For some applications, it is advantageous to base turbidity sensing on scattering (from relatively large particles) only. Such turbidimeters employ optics between the sample cell and the light sensor, which block the transmitted light and admit only *forward-scattered* light to the sensor; this blocking is not required in another type of turbidimeter in which *back-scattered* light is sensed.

Three special units of measurement are often used in turbidity measurement. One of these is the JTU (Jackson Turbidity Unit), derived from an early version of a turbidimeter, the *Jackson Candle Turbidimeter.* It consists of a flat-bottomed glass tube, graduated in JTU, below which a special candle is mounted. The sample liquid is slowly poured into the tube while visually observing the image of the candle flame from the top of the tube. A reading is then taken when the image disappears in a uniform glow (i.e., when the intensity of scattered light equals that of transmitted light). The scale is based on turbidity caused by a suspension of diatomaceous earth in distilled water. Another unit is the FTU (Formazin Turbidity Unit), which is based on a solution of a chemical mixture called Formazin in distilled water. For nephelometric turbidimeters (see below) a unit known as the NTU (Nephelometric Turbidity Unit) was developed and standardized by a U.S. governmental agency.

Nephelometry (see Figure 7-1c) is based on the sensing of light scattered at 90° from the incident light beam, or, in some cases, at 90° from the surface of a liquid illuminated by a light beam at a relatively shallow angle with the surface (a variant of this method has been used when the sample fluid is air). In some designs a second light sensor, placed at 180° with the first, is employed to increase the scattered-light detection capability. The transmitted light is either absorbed by a hood or sensed by a light sensor for reference purposes. Since light is scattered only when turbidity exists, nephelometers provide increasing signal amplitudes with increasing turbidity. A relatively short light path tends to increase the upper end of the turbidity measuring range.

Figure 7-3 illustrates the operating principle of a *transmittance turbidimeter.* The sample liquid fills the glass tube, which also provides the optical surface for the windows at the light source and light sensor (photocell) necessary to complete the transmission path. Attenuation of the light beam, due to scatter and absorption in the sample, is indicative of the suspended solids concentration of the sample, and the conditioned output signal is displayed in ppm. This particular design includes a motor-driven reciprocating piston within the glass tube; it draws in and later expels the sample while simultaneously wiping the optical surface of the sampling chamber. Similar turbidimeters are designed for in-line installation, some of them including provisions for making the flow laminar in the measuring region.

Back-scatter turbidimeters typically contain light source, light sensor, and optics for both in a single sensing head that is mounted so that it is in contact with the sampling fluid. The light sensor detects scattered and reflected light from the suspended solid particles in the sample fluid. In forward-

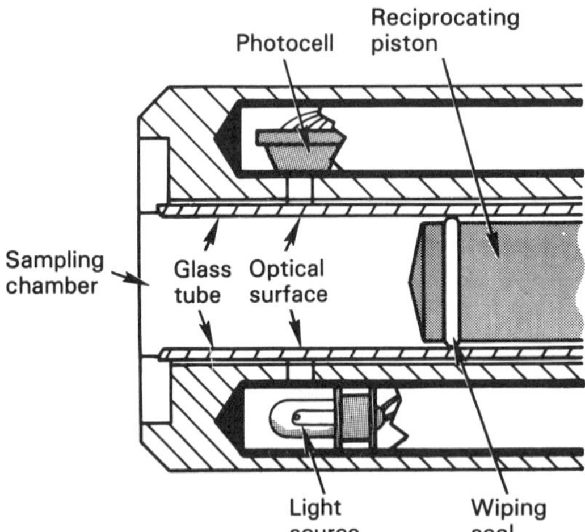

Figure 7-3. Transmittance turbidimeter (Courtesy of Inventive Systems, Inc.)

scatter turbidimeters, the light-sensor optics are so designed that the transmitted light is blocked and only forward-scattered light is detected. Alternatively, two light sources and two sensors can be employed in a pulsed mode; scattered and unscattered light is detected, and the ratios of those two light intensities can be computed by appropriate circuitry. An important design characteristic is freedom from error-causing effects of contaminating deposits on the walls of the sampling chamber when such deposits would be within the transmission path.

Nephelometric turbidimeters measure turbidity by measuring the light scattered by the sample fluid (usually a liquid) at 90° to the incident light beam. Additional scatter and even transmittance is measured in at least one type of design. As shown in Figure 7-4, the beam from a lamp, collimated by lenses, enters the sample cell (from the bottom, in this design) and passes through the sample fluid to a light shield. Light scattered from the sample at 90° is detected by the light sensor (a photomultiplier type in this unit) after passing through a window in the side of the cell holder assembly. In the laboratory-model nephelometric turbidimeter illustrated, one of several (Formazin dilution) sealed turbidity standards, or a focusing template, can be substituted for the sample cell. The turbidity standards are used to calibrate the instrument and the focusing template is used to adjust the position of the lamp. Circuitry is used to condition the photomultiplier output signal so that the readout on the indicating meter is in NTU (in five ranges, from 0 to 0.2 to 0 to 1000 NTU).

The nephelometric principle is also used in many on-line turbidimeters. Some examples of designs are shown in Figure 7-5. In the low-range nephelometric turbidimeter (Figure 7-5a), a strong light beam is passed through the stable, smooth surface created by the surface tension of a slowly flowing sample. With surface scattering minimized, sufficient 90° scatter is seen by the submerged photodetector to allow the measurement of very low turbidities, down to 0.001 NTU. A similar fluid-surface window is used in the Surface Scatter ® Turbidimeter (Figure 7-5b), in which a light beam focuses on a smooth fluid surface at an acute angle. One advantage of this design is that there is no contact between process stream and optical surfaces, minimizing the need for cleaning of the instrument. Most of the light striking the surface of the fluid is either reflected or refracted down the turbidimeter tube; a small amount is scattered by suspended particles and is incident upon a photodetector mounted directly over the point where the light enters the fluid, but at 90° to the light beam. Scattering occurs at or near the surface at high turbidities, but continues to a farther depth at low turbidities. This provides an inherent self-adjustment of path length and allows the measurement of low and very high turbidities.

The Ratio ® Turbidimeter (Figure 7-5c) employs three photodetectors,

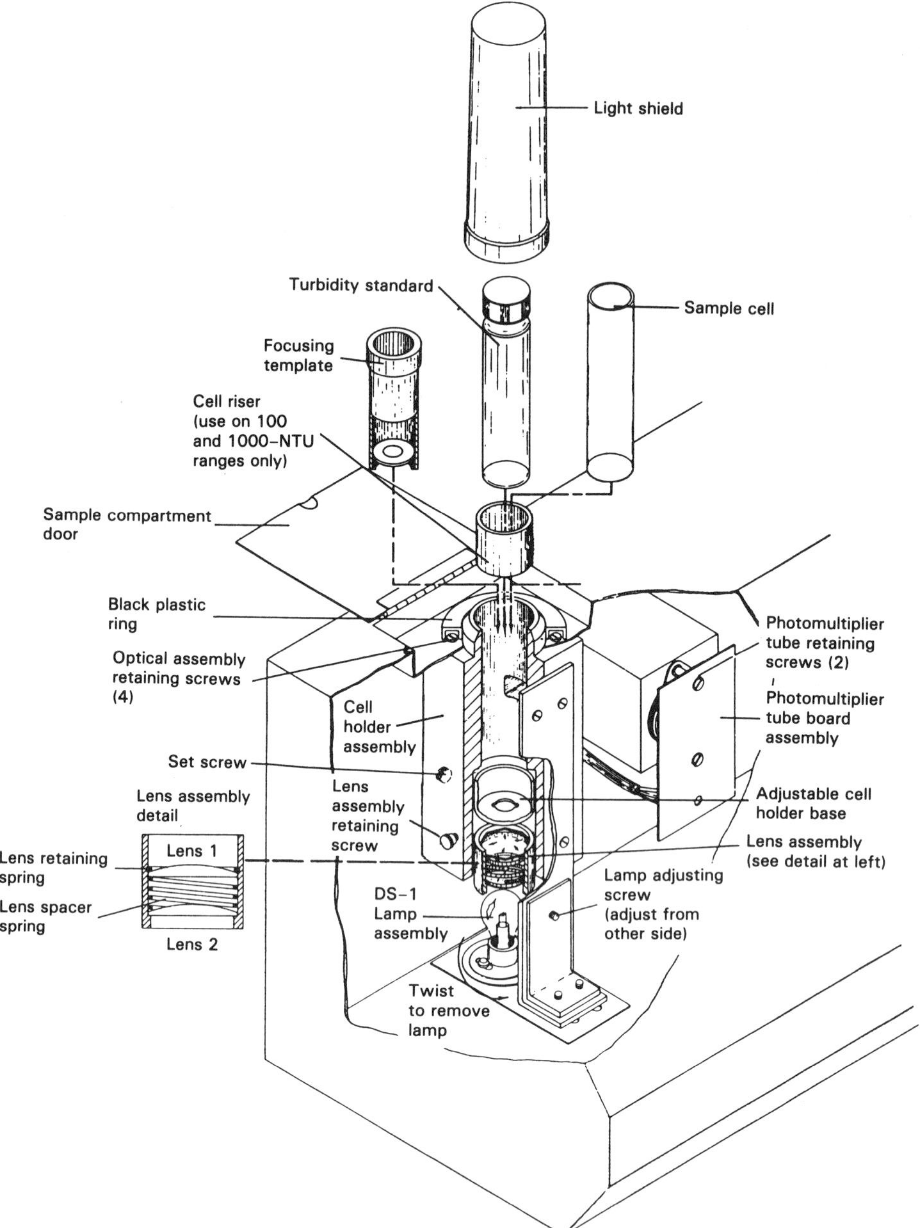

Figure 7-4. Optical assembly of laboratory-style nephelometric turbidimeter. (Courtesy of Hach Company.)

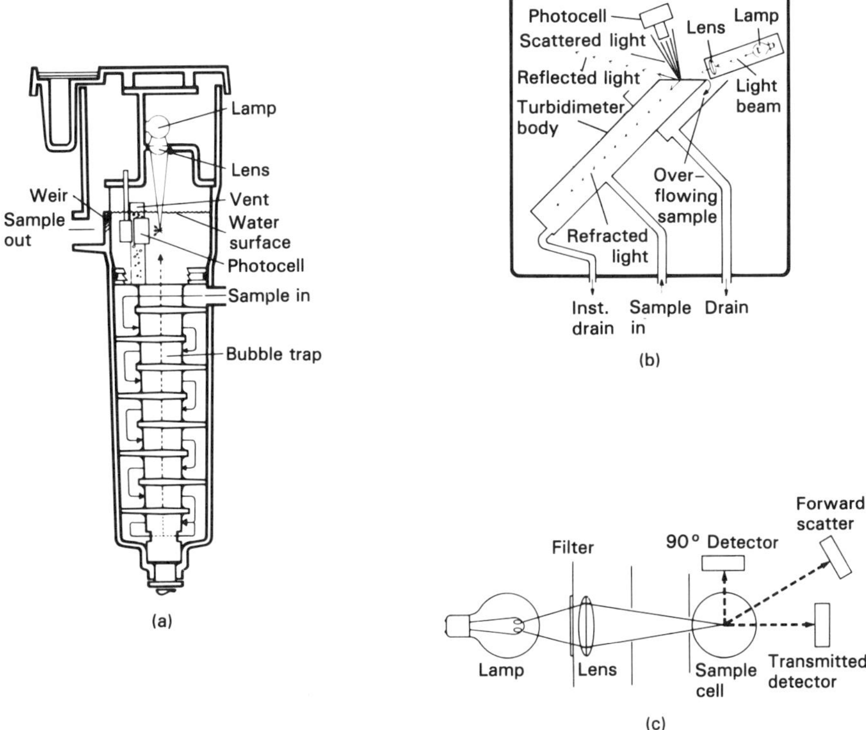

Figure 7-5. On-line nephelometric turbidimeters—schematic diagrams: (a) submerged photodetector design for low turbidity levels; (b) surface scatter turbidimeter; (c) ratio turbidimeter. (Courtesy of Hach Company.)

one for 90° scattered light, one for forward scattered light, and one for transmitted light. The output of this instrument, provided by built-in electronics, is the ratio of the conditioned signals from the 90° detector to a weighted sum of the outputs from the forward-scatter and transmittance detectors. This ratio method tends to compensate for absorption due to color in the sample, as well as for film build-up which affects each detector equally.

The main application of nephelometric turbidimeters is in water and waste-water analyses.

Nephelometers are, generally, nephelometric turbidimeters used for turbidity measurements in gaseous samples, especially for the detection of the amount of particulate matter suspended in air. The widely-used residential smoke detector is an example of a simple nephelometer which, however, activates an alarm rather than providing an analog output. Various types of nephelometers, detecting light scattered at angles not only at 90° but covering up to 180°, are used for atmospheric studies, including pollution and visibility determinations.

7.5 REFRACTOMETERS

Refractometry (see Figure 7-1e) is a measurement of the *index of refraction* of a substance. *Snell's law* states that as a light beam enters a more dense medium it will deflect (it *refracts*) toward the normal (with the direction of transmission). The index of refraction, *n*, is the ratio of the velocity of light in empty space to the velocity of light in a given material; it is a function of wavelength as well. Refractometry is usually based on sensing light reflected at the *critical angle*. When light through a more dense medium (whose index of refraction is n_2) meets the boundary with a less dense medium (whose index of refraction is n_1), the critical angle (ϕ) is the smallest angle with the normal (to the boundary) at which the light is totally reflected back into the more dense medium. The sine of the critical angle is equal to the ratio n_1/n_2. There are other methods of refractometry, such as the comparison of refraction loss through a sample to transmittance through a reference sample, by interferometric means in the *Rayleigh refractometer*.

Instruments which measure the index of refraction of substances are frequently used in process measurement to obtain not only refractive index measurements but determinations of solids concentration in liquids that can be correlated with index of refraction. A typical instrument is shown schematically in Figure 7-6a. It contains a light source, optics for beam collimations, a dual-element (photoconductive-junction) light sensor, and a prism which is brought in contact with the measured liquid. The prism in this design is made of spinel (synthetic diamond) since this material is very hard, resists abrasion and contamination, and has a long operating life. The light beam is directed at the boundary between prism and liquid. The light incident at this boundary changes sharply from mostly transmitted light to totally reflected light at the critical angle (see Figure 7-1e). The critical angle is dependent upon the index of refraction of the liquid and, hence, on the dissolved solids concentration in the liquid. The reflected-light image as seen by the dual-light-sensor (photodetector) is illustrated in Figure 7-6b. With increasing index of refraction (increasing solids concentration) the upper light sensor (sample detector) receives decreasing light, due to the change in critical angle, while the lower light sensor (reference detector) receives a constant amount of light. The two detectors are connected as a voltage divider with one leg of fixed resistance and the other leg of variable resistance so that an output voltage proportional to index of refraction is obtained (Figure 7-6c).

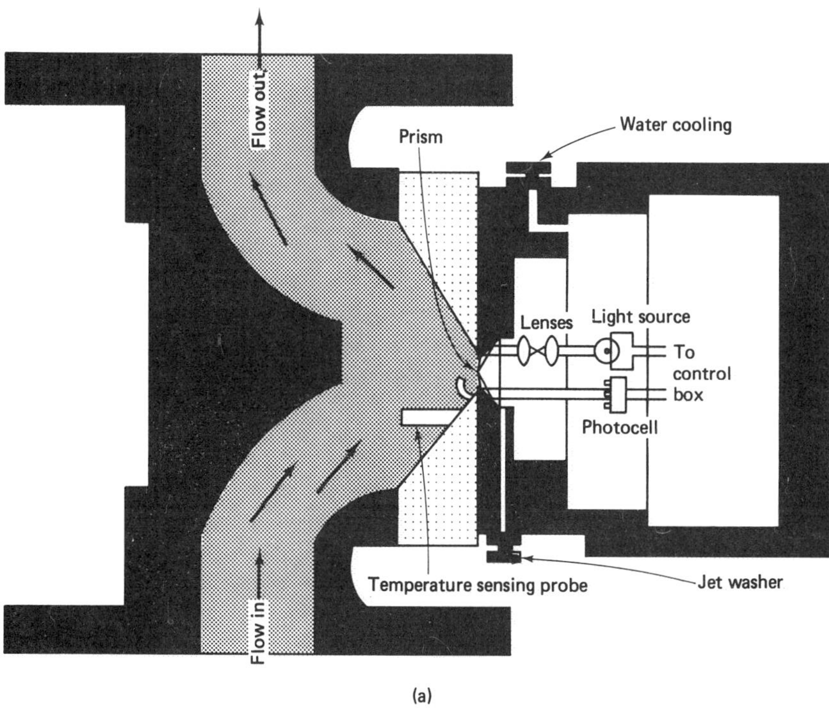

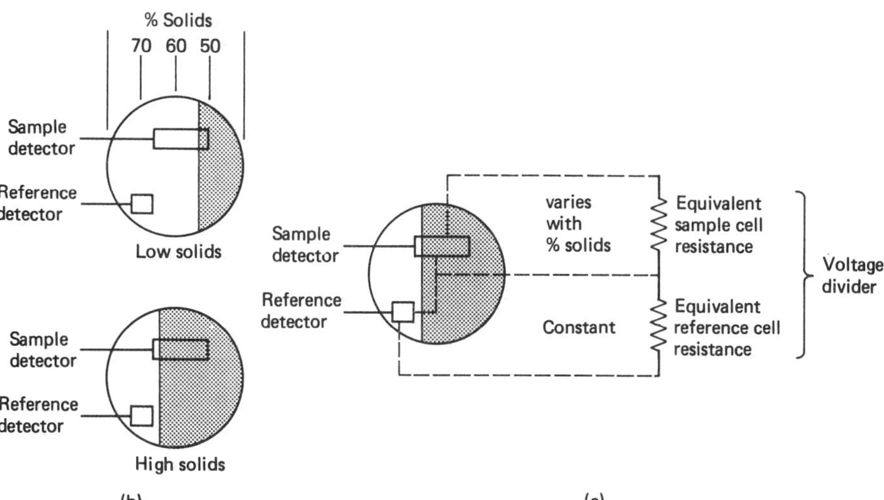

Figure 7-6. Operating principle of a process refractometer: (a) schematic layout of instrument assembly; (b) refractometer image; (c) simplified measuring circuit. (Courtesy of Anacon Corp.)

The instrument shown in Figure 7-6 is of the valve-body type, provided with flanges for mounting in a pipeline. An instrument of similar design but of the immersion-probe type, is shown in Figure 7-7; it is installed by threading it into a standard threaded fitting on a pipeline, tank, or other vessel. Both designs incorporate a temperature sensor and associated circuitry to compensate the refractometer readings for temperature as well as to provide a separate readout of temperature. Some of the other refractometer designs include, instead, a thermostatically controlled constant-temperature bath surrounding the sample line or cell (the index of refraction varies with temperature). The readout can be in terms of refractive index, of percent solids (in a specified liquid), or for sugar solutions, in Brix degrees, which are units of the hydrometric Brix scale.

Figure 7-7. Insertion-probe-type refractometer. (Courtesy of Anacon Corp.)

7.6 FLUORIMETERS

In *fluorimetry* (see Figure 7-1d), the light source emits a beam of light at wavelengths extending downward to about 200 nm (ultraviolet light). A filter (primary filter) is placed between light source and sample cell to enable the selection of the appropriate wavelengths. The fluorescent light from the sample is filtered to exclude nonfluorescent products and sensed at right angles to the incident beam. Fluorescence occurs when light is first absorbed, then reemitted from a substance after a very short time (about 10^{-8} s); when the elapsed time is between about 10^{-4} and over 5 s, the reemitted light is known as *phosphorescence*. It has been found that the fluorescence intensity from very weak solutions (around 11 ppm) is proportional to concentration. Though sometimes used for such weak concentrations, fluorimeters are much less commonly used than most other photometric instruments.

7.7 POLARIMETERS

Some substances can be analyzed by measuring the rotation of the plane of polarization of plane-polarized light when it passes through the sample. In its basic layout (see Figure 7-8), a polarimeter contains a light source whose light is linearly polarized by a device such as a Nicol prism. The polarized light is passed through the sample and then through an analyzer (e.g., another Nicol prism) to a photodetector. The sample substance must be optically active to a sufficient extent that the optical rotation of the substance can be used to provide a qualitative or quantitative analysis of the substance. Substances that can be analyzed by polarimetry include dextrose and sucrose.

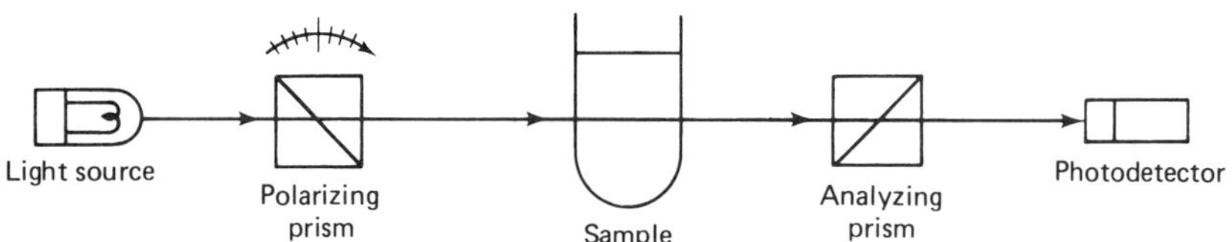

Figure 7-8. Polarimeter—basic schematic arrangement.

Various versions of electronic polarimeters (photopolarimeters) have been developed; a typical method of operation involves obtaining a null balance by rotating the polarizing prism until it matches the optical rotation produced by the sample and reading the polarizer angle when a null condition is achieved. A polarimeter becomes a *spectropolarimeter* by the addition of a scanning or adjustable monochromator between light source and polarizer. In a remote-sensing version of a spectropolarimeter, relatively coarse determinations of polarization of incident light at several different wavelengths are made by using two filter wheels, one containing wavelength filters, the other containing polarizing filters, with the wheels so positioned that a filter of one wheel is always used in conjunction with a filter of the other wheel.

7.8 CHEMILUMINESCENCE ANALYZERS

Certain compounds (molecules) can be detected by mixing them with a reagent with which the molecules react to form excited molecules which decay spontaneously with photon emission. This emission (see Figure 7-1f) is then passed through an optical filter (or filters) and sensed by a photodetector. Among the reactions that have been used for gas analyses are the chemiluminescent reaction of ozone (O_3) with nitric oxide (NO); either gas can be detected by admixture of a controlled amount of the other; additionally, NO_2 can be detected by converting it to NO. Another reaction resulting in chemiluminescence is that of ozone with ethylene; this is used for O_3 analysis.

A major application of chemiluminescence analyzers is in NOx (oxides of nitrogen) analysis, for example in stack gas emissions. A typical design for such an application provides four measuring ranges (0 to 10, 100, 1000, 10 000 ppm NOx), zero and span drifts of 1% in 24 hours, 1% linearity, and a 90% response time of 1 s. The signal output is a dc voltage, or, optionally, a dc current. Additional alarm outputs are also optional. Power requirements are 1 kW at 115 V ac, 50-60 Hz. Shielding is provided for all high voltages.

7.9 FLAME PHOTOMETERS

Flame (emission) photometry is a form of emission spectrophotometry (see Section 9.1) wherein atoms and molecules are raised from a ground state to an excited (electronic) state by thermal collisions with flame constituents and then emit characteristic electromagnetic radiations while they return to their ground state.

The sample to be analyzed is first converted into a vapor by means of an atomizer (nebulizer), which converts the sample (if liquid) into a fine vaporlike mist. The sample is then fed into the capillary of the burner (see Figure 7-9), where it is burned in a fuel–oxidant mixture. Typical burner mixtures are hydrogen–oxygen, acetylene–air, acetylene–oxygen, and acetylene–nitrous oxide.

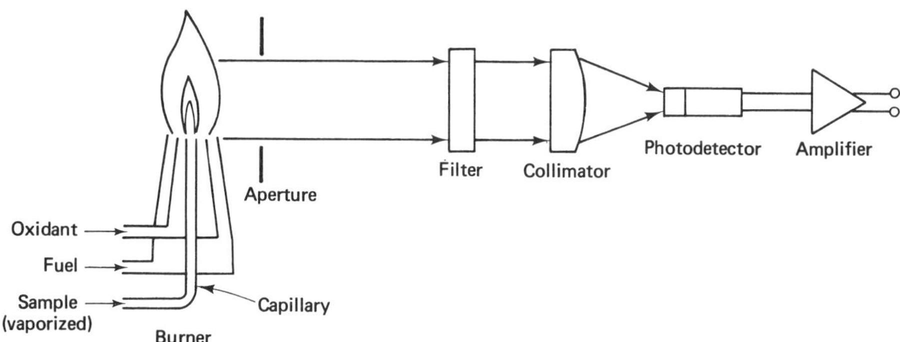

Figure 7-9. Flame photometer—basic schematic arrangement.

The light emissions from the sample in the flame are passed through a narrow-band-pass filter and then collimated onto a photodetector. The filter is selected for the wavelength of maximum emission of the substance whose relative abundance or concentration is to be determined. The amplified output of the photodetector can then be displayed on a meter, a strip-chart recorder, or digital printer. The use of the simple flame photometer is limited to such substances as those containing metal ions that provide a sufficiently bright emission, at a wavelength in the visible or near-visible portion of the spectrum, for a reasonably high signal-to-noise ratio. Much wider applications exist for the related *flame emission spectrometer,* in which the single band-pass filter is replaced by a scanning monochromator (see Section 9.5).

Related to the flame photometer as well as the emission spectrophotometer is the *spark-emission UV photometer,* which is used for the detection of halides, especially of the presence and abundance of halogen compounds in the air. The air is sampled and the sample is fed to a chamber in which an electrical spark is generated. The emission from the spark is enhanced in the ultraviolet region by the halogens and the brightness in that portion of the spectrum is indicative of halogen concentration. The light emission from the spark chamber is passed through a UV filter and then to a photodetector.

The instruments described above are in a category that has recently become known as *nondispersive photometric analyzers.* They differ from spectrophotometers and IR spectroradiometers in that they lack the scanning monochromator which disperses the polychromatic light into a set of monochromatic beams, each of which is then analyzed for intensity, usually sequentially over one scan, by a photodetector. Nondispersive analyzers employ, instead, a narrow-band-pass filter selected for the wavelength of maximum emission (or of maximum absorption) of a specific substance whose presence/absence or relative abundance or concentration in a mixture is to be determined.

7.10 NONDISPERSIVE INFRARED ANALYZERS

A *nondispersive infrared analyzer* (*NDIR analyzer*) is an absorption spectrometer in which the scanning monochromator is replaced by one (sometimes two) narrow-band-pass filters and which operates in the near or middle infrared region of the spectrum. The operation in the IR region dictates the use of an IR source (instead of a visible or UV source), an IR detector, and good IR characteristics of all internal optical elements.

The sample cell and its length are also important. Absorbance in the sample cell, as given by the Beer-Lambert law, is the product of cell path length, the absorbtivity of the sample at a given wavelength, and the concentration of the component of interest. When the concentration of a gas in the ambient air is to be analyzed, the sample cell is sometimes omitted and the optical path (whose length is still critical) is exposed directly to the ambient air. Some NDIR analyzers use multiple internal reflections (MIR) to obtain a large increase in the optical path length without significantly increasing the size of the instrument.

The IR region of the electromagnetic spectrum has been found particularly suitable for absorption analyses of fluids, primarily of gases, although a fair number of designs can also analyze liquids. About 250 different gases have usable absorption lines in the IR region. Some models are specifically designed to monitor such gases as CO, CO_2 and combustible gases in air.

NDIR analyzers exist in many different optical configurations. The basic configuration, however, is one of three types: single-beam, dual-wavelength; dual-beam, single-wavelength; and dual-beam, dual-wavelength.

A typical layout of a dual-beam, single-wavelength analyzer, designed primarily for gas analysis, is shown in Figure 7-10. IR emission from a source is collimated into two beams. A chopper disk alternatingly enables one beam and disables the other; it also provides synchronization signals. One beam passes through a cell containing (or exposed to) the sample being analyzed; the other beam passes through a cell containing a reference gas. The two

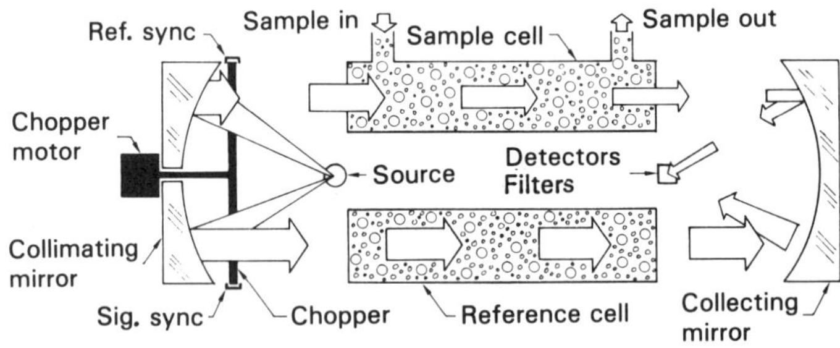

Figure 7-10. Typical schematic layout of a dual-beam, single wavelength NDIR analyzer. (Courtesy of OFC Infrared Instruments, Div. of Optical Filter Corp.)

beams exiting their respective cells are collected and collimated onto an IR detector through a filter selected to pass radiation at a wavelength corresponding to the wavelength of maximum absorption in the IR portion of the spectrum. The photodetector output alternatingly represents a background level, established by the reference beam, and a measuring level, established by the sample beam. The difference between those two outputs is indicative of the concentration, in the sample, of the gas of interest. When two gases in the sample are to be analyzed and their wavelengths of maximum absorption are far enough apart, two filter-photodetector sets can be employed for the concentration analyses of both gases.

Several design approaches have been used to produce a dual-beam, dual-wavelength NDIR analyzer. One design uses separate source-filter sets to provide a measurement beam and a reference beam. Both beams pass through the sample chamber through a window and are reflected at the opposite side of the chamber by a mirror. The reflected beams then pass back through the window and are incident on a measuring detector and a reference detector, respectively.

In single-beam, dual-wavelength NDIR analyzers, different techniques are used to produce the two wavelengths. One example is a layout in which a polychromatic beam from an IR source passes through the sample and is separated into two beams by a beam splitter. The two beams are incident upon two filter-detector sets, with one narrow-bandpass filter selected for the wavelength at which maximum absorption occurs; the other is selected for a reference wavelength unaffected by the absorption characteristics of the sample.

Another design (see Figure 7-11), usable for gas as well as liquid samples,

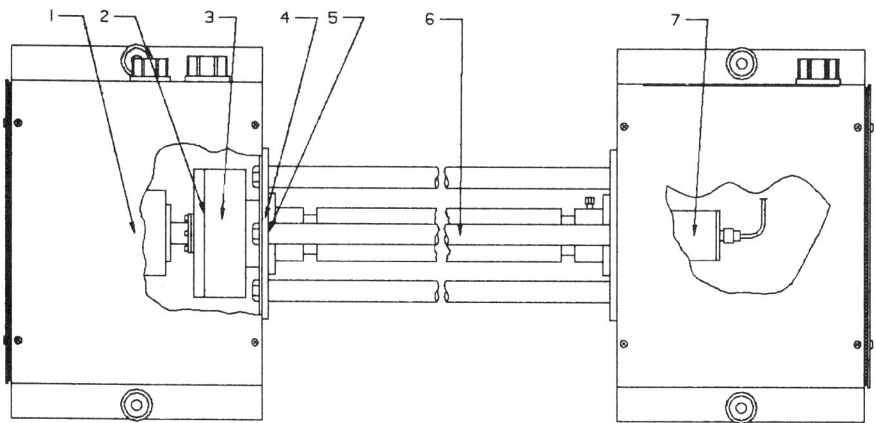

Figure 7-11. Single-beam, dual-wavelength NDIR analyzer using filter wheel: 1—source; 2—chopper motor; 3—filter wheel assembly; 4—lenses; 5—cell windows (one at each end of cell); 6—sample cell; 7—detector. (Courtesy of ABB Process Analytics.)

has a filter wheel containing two measuring and two reference filters. Two IR sources are available for this instrument: a tungsten lamp for near IR (0.75 to 2.5 μm) or a platinum source for the wavelength band from 2.5 to 15 μm. A choice of six different sample-cell window materials is similarly wavelength-dependent. Several cell- and seal-material combinations are offered, depending on the type of sample fluid. Different path lengths are offered, depending on type of sample fluid (gas, liquid, or acqueous) and the concentration and wavelengths ranges. The instrument employs a lithium tantalate pyroelectric detector.

The instrument illustrated in Figure 7-12 is specifically designed for the NDIR analysis of combustible gases in the ambient air and, therefore, provides an open sample cell. It is also of the single-beam, dual-wavelength type. A mirror reflects the beam from the source back through the sample (air) and focuses it on two narrow-band interference filters, one sensitive to the selected absorption wavelength of hydrocarbons, the other used as reference. A

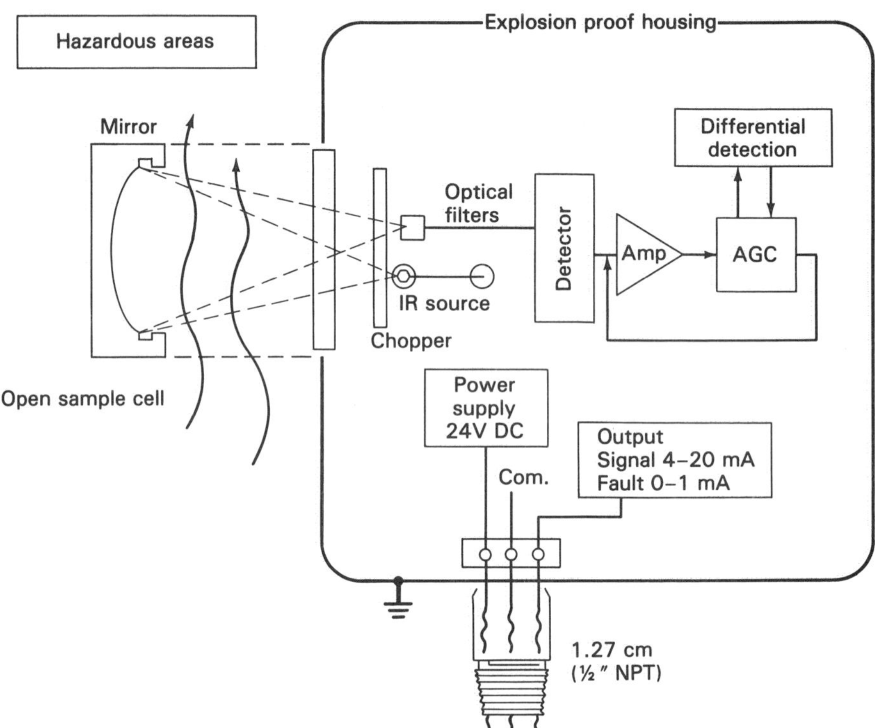

Figure 7-12. Single-beam, dual-wavelength NDIR analyzer, for combustible gas detection in air, using fixed-position filters. (Courtesy of Astro International Corp.)

chopper interrupts the beam so that the detector alternatingly sees measurement and zero. Electronics extract the ratio between measurement and reference signals. The output of the instrument is 4–20 mA dc; additionally, a system fault is indicated by 0–1 mA dc. The absence of a closed sample cell helps the instrument achieve a short response time, 3 seconds or less.

7.11 NONDISPERSIVE ULTRAVIOLET ANALYZERS

NDUV analyzers are useful for concentration measurements of certain compounds such as H_2S, SO_2, and Cl_2, as well as phenols in acqueous solutions; they can also be used for NO and NOx analyses.

One design is very similar to the single-beam, dual-wavelength analyzer shown in Figure 7-11; it differs mainly in the choice of source and detector. A deuterium lamp is used as source, and a silicon photodiode is used as detector. Its wavelength range is 200–800 nm.

A different design is shown in Figure 7-13. This gas analyzer uses the

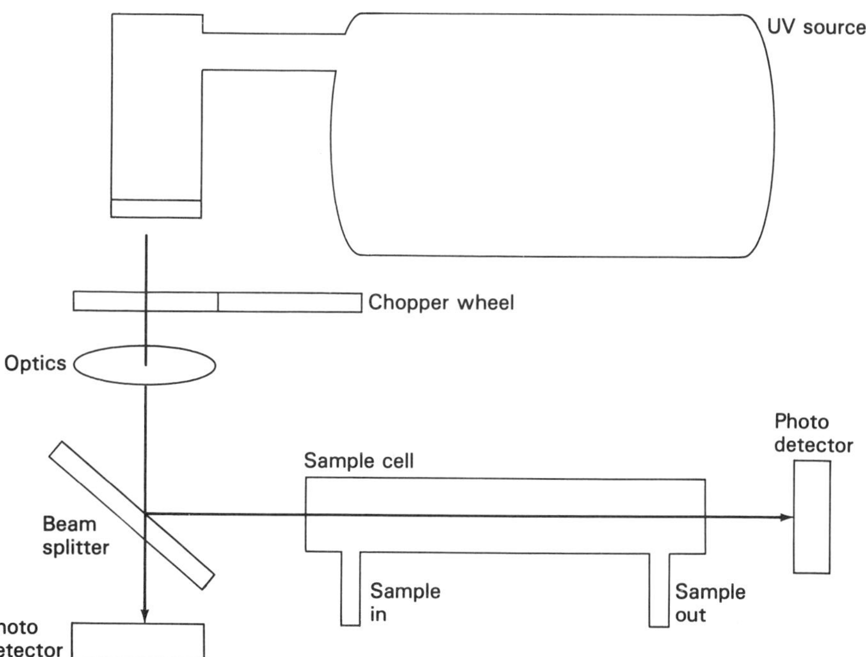

Figure 7-13. Dual-beam, single-wavelength NDUV analyzer. (Courtesy of Applied Automation/Hartmann & Braun.)

dual-beam, single-wavelength optical configuration. It operates over the UV spectral region of 210–270 nm. The source is a hollow-cathode lamp. The beam from the source is passed through a chopper wheel which contains a sealed reference-gas cell. The beam is then divided by a beam splitter. The chopper wheel provides time separation between the reference and measurement beams. The reference beam and the measuring beam (which passes through the sample cell) each use their own solid-state photodetector. The source is of a special design and contains an extra large reservoir for reactant components. The optical section and the electronics section are each contained on slide-out trays in separate housings. The electronics tray is also hinged for ease of maintenance, field tests, and calibration. The manufacturer also provides for proper sample conditioning as required by the application (e.g., dust filtering, pressure and flow-rate regulation, and cooling). Instrument output signals are dc current (0–20 and 4–20 mA) and dc voltage (0–10 or 2–10 V), selectable by the user.

8

CHROMATOGRAPHS

Chromatographs are used for the qualitative and quantitative analyses of mixtures of chemical compounds. Their operation is based on the sequential separation of compounds in a tube containing an adsorbent. There are two basic categories of chromatographs: gas chromatographs (GC) and liquid chromatographs (LC). GCs have been in development longer than LCs and there are many more of the former instruments in use than the latter. Most analyses can be accomplished by a GC; however, some types of analysis either present problems when a GC is used, or simply cannot be accomplished with a GC. The LC, then, provides additional analysis capabilities. Both categories are available as laboratory instruments or as on-line analyzers, again with process GCs at a higher level of development than process LCs.

8.1 GAS CHROMATOGRAPHS

A basic schematic of a gas chromatograph is shown in Figure 8-1. The sample is injected into a stream of carrier gas, which carries it through a *column* (separation column), a long, thin tube, typically in the form of a multiturn coil, which contains a specific adsorption material (the *packing*). Substances

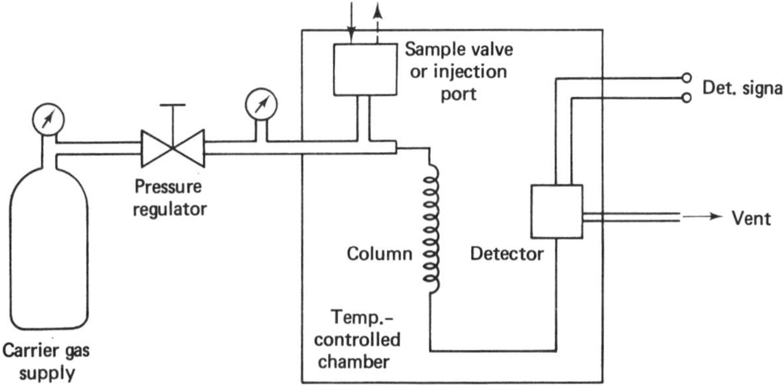

Figure 8-1. Gas chromatograph—basic schematic.

with a high affinity relative to the packing travel through the column at a slower rate than substances with a low affinity. The compounds exit from the column in an identifiable sequence, due to the *elution* process in the column. The elution products, still carried by the carrier gas, pass through a detector, which provides an output signal indicative of the abundance of each compound. After detection the effuent is either vented or collected.

The *carrier gas,* usually hydrogen, nitrogen, or helium, is either stored in a pressure bottle or (as is sometimes the case with hydrogen) generated in situ. A pressure regulator adjusts the pressure and flow rate of the carrier gas through the system. In some designs the flow rate is monitored by a gas flowmeter.

The sample valve or injection port, the column, and the detector are all mounted within a *temperature chamber.* A proportional temperature controller maintains the electrically heated chamber as well as its components, at a selectable temperature. For some analyses it is required that the temperature be raised at a special linear rate (*programmed-temperature chromatography*). Instrument designs capable of providing such a function are equipped with a programmer in the heater control electronics.

In process chromatographs, a closely controlled quantity of the sample is introduced into the GC by means of a *sample valve.* Sample valve design and operation is critical to obtaining good analyses; the process is sampled by the valve many times during a day and the number of molecules in the sample must be repeatable. Liquid samples are vaporized upon entering the GC, either by a separate electrically heated vaporizer or (when the temperature chamber is sufficiently hot) in the sample valve or port itself. In laboratory GCs, the sample is usually introduced by a *microsyringe* through a self-sealing rubber *septum.* The sample gas or vapor is thus introduced into the carrier gas as a "plug" and carried into the column.

When solids are to be analyzed they are first converted into a vapor in a *pyrolizer* oven. Depending on the type of analysis, the pyrolizer is either temperature-controlled with the temperature increased in discrete and known steps, or the temperature is raised linearly with time, at a selectable rate, which can be quite rapid for some analyses. The pyrolysis is compound-selective since different substances will volatilize at different temperatures. The temperature of the pyrolizer oven is monitored closely by a temperature sensor with fast response and the temperature readings are correlated with the GC detector readings for data interpretation.

There are two basic types of columns. *Packed columns* are coiled or U-shaped tubes with an ID of 2 to 5 (sometimes up to 10) mm, which are packed with a solid material (*solid support*) that is then coated with a selected solvent referred to as the *stationary phase* (as opposed to the *mobile phase* constituted by the carrier gas, carrying the sample). Such columns are used in *gas–liquid chromatography* since the stationary phase is a liquid phase. *Capillary columns* are long, thin (typically 0.1 to 0.5 mm ID) tubes whose inside surface is coated with, usually, a liquid phase, sometimes liquid on a very fine mesh solid support, sometimes a solid absorbent (*gas–solid chromatography*). Packed columns with an ID down to 0.5 mm are used sometimes. Capillary columns are usually coiled. The tubing material can be glass, metal (stainless steel, copper, aluminum), which may have an internal Teflon coating, or plastic (Teflon, nylon, polyethylene). A large variety of solid supports are used in packings, including textured glass beads, plastic powders, silica gel, activated charcoal, and activated alumina. An even larger variety of stationary phases are available; they are usually identified by trade names proprietary to a particular manufacturer; examples are Chromosorb 101, Porapak Q, Apiezon L, Carbowax 20M, OV-210, and DEXSIL 300GC. The stationary phase must be carefully selected for a given analytical problem, with emphasis on its separating ability and thermal stability.

Detectors used in GCs are described in other chapters of this book. The four most commonly used detectors are the flame ionization detector (FID), the photoionization detector (PID) (see Section 6.2), the flame photometer (see Section 7.9), and the thermal conductivity cell (TCC) (see Section 5.2). Some GCs use two detectors in tandem; some others may use a separate detector which responds only to the carrier gas. Use of a TCC requires that the carrier gas be supplied to its two reference filaments; in some designs the carrier gas travels through a reference column before it reaches the TCC.

The detector output is typically a very low-level signal, which is amplified and usually recorded on a strip-chart recorder for data interpretation. An example of a chromatogram (for a test mixture) is shown in Figure 8-2. Data interpretation involves determination of the peak height and peak area. The latter can be calculated coarsely by multiplying the peak width at one-half

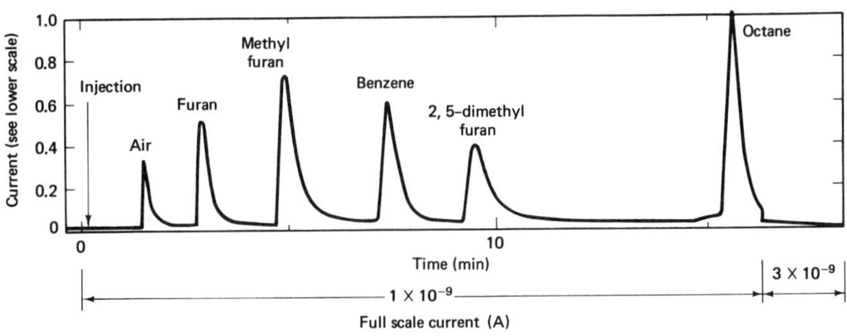

Figure 8-2. Example of a chromatogram (ionization detector output as fraction of full-scale output current).

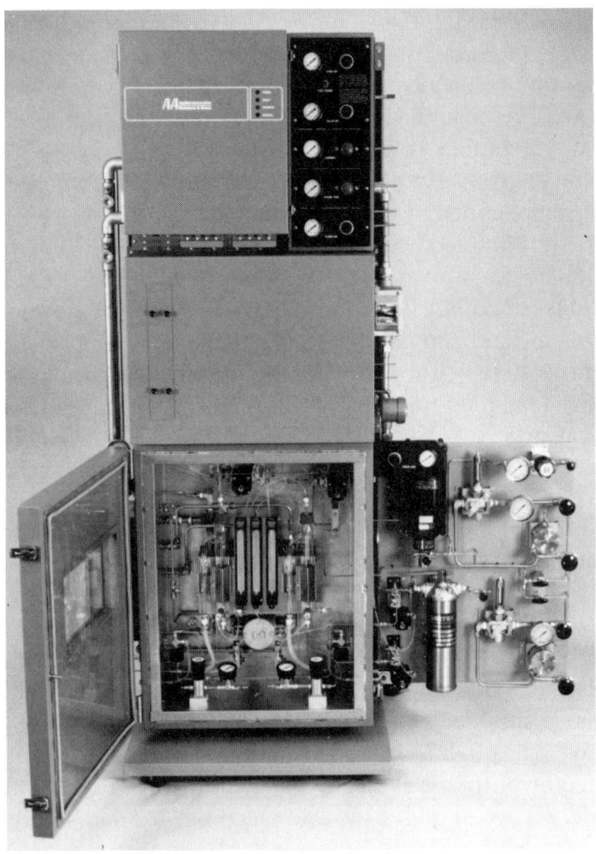

Figure 8-3. Process gas chromatograph (with sparging capability). (Courtesy of Applied Automation/Hartmann & Braun.)

peak height by one-half the peak height. More accurate results are obtained by integrating the peak-area, an operation that is greatly facilitated by use of a microprocessor. Most modern GCs incorporate microprocessors that can perform the required peak-analysis operations in addition to performing other functions, such as control operations.

An example of a process gas chromatograph design is illustrated in Figure 8-3. The design shown has a sparging unit added for enriching the sample. This unit consists of a borosilicate glass sparger, pressurization vessels, sample heating equipment, and flow controller. This instrument is specifically designed for the analysis of volatile contaminants in water and waste-water. However, the illustration is representative of other general-purpose process GCs. The top portion is the microprocessor-based controller-analyzer. The center portion is the temperature-controlled analysis section, which can be equipped with packed, micropacked, or capillary columns and any of the four detectors mentioned above. The oven temperature is software selectable. The bottom portion (door open) shows the sample system.

A portable GC is shown in Figure 8-4. It relies on sample injection, optionally by syringe; otherwise automated sampling is offered, using six

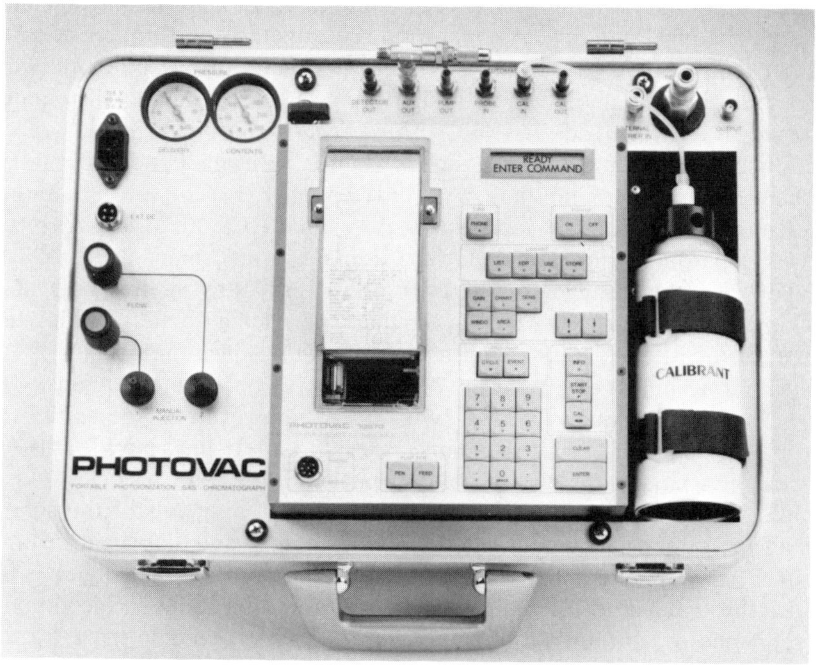

Figure 8-4. Portable gas chromatograph. (Courtesy of Photovac Inc.)

three-port miniature valves. Air or an inert gas is used as carrier, and the detector is of the photoionization type. The instrument incorporates a pre-column as well as a (much longer) regular column, both made of fused quartz (use of the pre-column simplifies back-flush). This portable instrument is a complete GC system, which, with a mass of 13.3 kg and dimensions of 46 × 16 × 34 cm, incorporates a multicolor strip-chart recorder, calibration gas, microprocessor-based control and peak analysis, and a communications port. The recorder can print out compounds by name and concentration, based on stored libraries. The unit is powered by internal batteries, external batteries or ac adapter (required for the isothermal oven option), or ac line power (115 or 230 V). The analysis time is typically somewhat less than ten minutes. Very thorough analyses are made possible by combining a GC with a mass spectrometer (see Chapter 11); this combination is referred to as GCMS.

8.2 LIQUID CHROMATOGRAPHS

Several techniques are used in liquid chromatography. One factor they have in common is the absence of a carrier gas. In *liquid-liquid chromatographs* (*LLC*), a liquid stationary phase coats solid packing material in the column through chemical bonding. A separation is attained by partitioning of a sample between a liquid carrier and the liquid stationary phase. When the liquid carrier is more polar than the stationary phase, the technique is frequently referred to as "reversed-phase LC". *Ion-exchange chromatographs* separate the ionic components in an aqueous liquid by using bonded-phase ion-exchange resins as column packing. This technique is useful for obtaining the sequential separation of inorganic compounds. *Size-exclusion chromatographs* separate components of the sample by molecular size: molecules smaller than the particles of the packing material stay in the pores of the packing longer than molecules larger than the packing material. This technique, also called *gel permeation chromatography,* since the packing is typically a porous gel, is useful for the on-line analysis of polymer streams.

The most frequently used type of LC, however, is the *liquid-solid chromatograph*. Solvent, from a tank, is used to dilute the sample. This solution is then passed through a column packed with solid particles. As the sample solution elutes through the column, the components of the solution will adsorb on the packing for varying periods of time and thus separate on the basis of travel time through the column. The effluent from the column is then passed through a detector, which provides information about each component in sequence of elution. After passing through the detector the effluent is vented or collected.

Detectors used in LCs are quite different from the gas detectors used in

GCs. They are usually photometric in nature. One popular type of detector is the optical adsorbance photometer, which operates either in the UV region or over the wider UV-to-near IR region; it is related to the nondispersive analyzer (see Section 7.11) and uses separate reference and sample flows. Another frequently used detector is the refractive index detector, related to the refractometer (see Section 7.5). A relatively new type is a capacitive detector which responds to the variations of the dielectric constant of sample components.

In many current designs, particularly in-line LCs, the sample solution is pressurized to a high pressure before entering the column, to speed up the analysis. The term high-pressure liquid chromatograph (HPLC) was often applied in such cases. Around 1970 the meaning of the initials "HP" was changed to "high performance."

LCs are sometimes combined with a mass spectrometer (see Chapter 11). In such an *LCMS,* the sample solution is carried into the mass spectrometer in vaporized form.

9

SPECTRORADIOMETRIC ANALYSIS INSTRUMENTS

9.1 SPECTROMETRY USING ELECTROMAGNETIC RADIATION

This category of instruments is used to determine chemical composition from the spectral distribution of electromagnetic energy. The electromagnetic radiation that is analyzed can be in the spectral regions of gamma rays, X rays, extreme (vacuum) ultraviolet, ultraviolet (UV), visible light, near-infrared (IR), far-infrared (see Figure 1-1 for all these regions), or submillimeter waves or microwaves (see Figure 10-2). The electromagnetic radiation whose spectral characteristics are analyzed can be due to emission from a hot sample (*emission spectrometry*) or to fluorescence, phosphorescence, or luminescence from a sample. Emission spectrometry is also used in astrophysics, where the hot "sample" is a nebula, galaxy, star, planet, moon, or comet. In *absorption spectrometry* a portion of the spectrum is scanned to determine at what wavelengths what quantity of radiation from a source is absorbed. The emission or absorption of electromagnetic radiation at specific wavelengths is due primarily to changes in electronic energy levels of atoms and electronic states of molecules and in changes (transitions) of molecular vibrational and rotational states.

An emission spectrum will contain a number of peaks at certain wavelengths, whereas an absorption spectrum will show a number of valleys at certain wavelengths. The wavelengths at which such peaks or valleys occur permit identification of an atom or molecule. The peak or valley amplitudes are related to the abundance or concentration (in a mixture) of the atom or molecule. The peaks and valleys (valleys, in absorption spectra, become peaks if the spectrum is shown in terms of absorbance instead of transmission) are generally very narrow and are called *spectral lines.* A group of closely adjacent lines is called a *spectral band.* Spectra of essentially all known atoms and molecules have been determined and are available, for correlation with experimentally determined spectra, in graphical or tabular form.

It is customary to express spectral distribution (location of spectral lines) in terms of photon energy when the wavelengths in the spectrum are shorter than that of ultraviolet. When the wavelengths are between those of ultraviolet and far infrared the distribution is expressed either in wavelength (nanometers or micrometers) or in wavenumber (the reciprocal of wavelength, expressed in cm^{-1}). In the microwave region the spectral distribution is usually expressed in terms of frequency.

As the *spectral resolution* of spectroradiometers improved it became possible to discern lines. With further improvements it was discovered that some of the lines were really groups of thinner lines (*fine* spectra) and with still further improvements it was shown that even some fine lines consisted of several extremely narrow lines (*hyperfine* spectra). The definition of spectral resolution is: the smallest difference in wavelength (or in photon energy for the case of X rays and gamma rays) for which separation of two adjacent lines is still possible. Spectral resolution can be stated in terms of $\Delta\lambda/\lambda$, where λ is wavelength, or in terms of wavenumber, or in terms of *resolving power,* $\lambda/\Delta\lambda$; the resolving power of spectrometers is generally between 5000, for simple instruments, and over 200 000, for spectrometers employing diffraction gratings. A resolving power of 200 000 means that two lines, separated by 0.0025 nm, can be resolved at 500 nm.

Analytical methods, including sample preparation, for spectrometry will not be described here. Several good texts are listed in the bibliography for this chapter. However, it can be noted briefly that absorption spectrometry involves the interaction of one or more beams of monochromatic radiation with the sample; emission spectroscopy involves heating the sample electrically, or by use of a flame, or by incorporating the sample material into an electrode which is then heated by causing a dc or ac arc, or spark, to be generated from this electrode; heating of gaseous samples can be attained in a discharge tube; fluorescence is short-lived light emission (10^{-8} to 10^{-3} s in duration) from a sample in response to incident electromagnetic radiation (typically UV; X-ray fluorescence is treated separately, in this chapter); phosphorescence is long-lived light emission (longer, typically much longer, than 1 ms); and chemiluminescence is light emission, caused by a chemical reaction, that persists still longer.

9.2 UV, VISIBLE-LIGHT, AND IR SPECTROPHOTOMETERS

9.2.1 Spectrophotometer Fundamentals

Spectrophometers are primarily those instruments in which the spectral distribution of radiant energy is measured in the spectral regions, including the UV, visible, and near-to-middle IR. Instruments that are similar in their operating principle to spectrophotometers are used in the extreme ultraviolet (XUV) and far-infrared (FIR) regions. The spectral regions are delineated as follows.

	Wavelengths (nm)	Wavenumbers (cm^{-1})
XUV (extreme ultraviolet, vacuum ultraviolet)	10–200	1 000 000–50 000
UV (ultraviolet)	200–380	50 000–26 315
VIS (visible light)	380–780	26 315–12 820
NIR (near-infrared)	780–3000	12 820–3333
MIR (middle-infrared)	3000–15 000 (3–15 μm)	3333–666
FIR (far-infrared)	15×10^3–8×10^5 (15–800 μm)	666–12.5

The MIR region is customarily included in the FIR region; however, in spectrophotometry it is useful to separate these two regions because 15 μm is approximately the upper limit of (quantum) detectors; thermal detectors have to be used for IR detection above this limit.

The three most commonly used types of spectrophotometers are illustrated schematically in Figure 9-1, in their most elementary form. The most essential element is the *monochromator,* whose optical input is polychromatic light and whose optical output is monochromatic light of known wavelength. Concentration and geometric delineation of the light beams are accomplished by *collimators* (typically lenses or concave mirrors) and by optical *slits* (entrance and exit slits for the monochromator). Radiation (light) at the wavelength of interest is sensed by a *photodetector,* which produces an electrical output signal proportional, in amplitude, to the light intensity at that wavelength.

In the basic *absorption spectrophotometer* (Figure 9-1a), light, from a

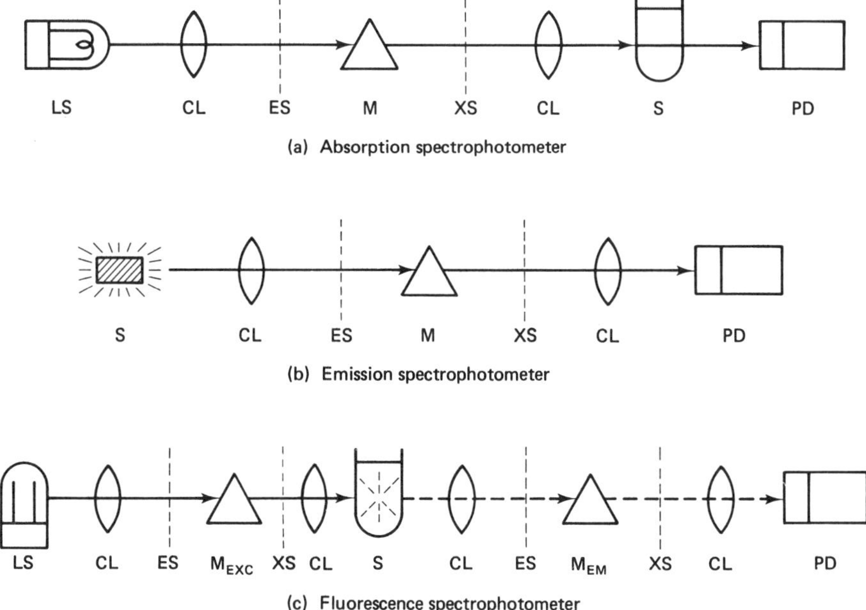

Figure 9-1. Basic elements of spectrophotometers: LS, light source; CL, collimator; ES, entrance slit; M, monochromator; XS, exit slit; S, sample; PD, photodetector; EXC, excitation; EM, emission. (a) Absorption spectrophotometer; (b) emission spectrophotometer; (c) fluorescence spectrophotometer.

source, is converted into monochromatic light, whose wavelength is adjustable and always known, which passes through the sample. Absorption will occur in the sample and the corresponding light changes are sensed by the photodetector. The absorption spectrum is then established by a plot of light intensity vs. wavelength. The spectral characteristics of the light source, photodetector, and all internal optical elements must be known and considered in the final readings.

The basic *emission spectrophotometer* differs from the basic absorption spectrophotometer primarily in the absence of a light source and the location of the sample. Polychromatic light from the (heated) sample is directed at the monochromator, which sweeps over the specified spectral region and then provides monochromatic light, at known wavelengths and intensities indicative of the emission spectrum, to the photodetector (see Figure 9-1b).

In the *fluorescence spectrophotometer,* monochromatic light wavelengths are selected from the polychromatic emissions of a light source (which emits in a spectral region, including UV). The monochromatic light from the *excitation monochromator* is directed at the sample, in which fluorescence will occur at certain excitation wavelengths. The fluorescent light, which is polychromatic, is then analyzed for spectral content by the *emission monochromator* in conjunction with the photodetector. This is illustrated schematically in Figure 9-1c.

It should be noted that the three schematics of Figure 9-1 are intended primarily to show the similarities and the differences between the three types of instruments and show only the most basic elements. An additional set of elements is common to virtually all spectrophotometers: a reference channel whose beam either does not pass through the sample or does pass through it but is detected by a photodetector through a filter that passes only a sample-indifferent band of wavelengths.

9.2.2 Monochromators

The most essential part of a spectrophotometer is its light-*dispersing* device, the *monochromator,* which permits light intensity at selected wavelengths, or in narrow bands of wavelengths, to be examined. The monochromator receives polychromatic light through an entrance slit, modifies this radiation geometrically and optically, and then leads selected radiation through an exit slit so that it can be sensed by a photodetector.

Optical *filters* are used in the simplest spectrophotometers. It should be noted, however, that filters may be more suitable than other types of monochromators in XUV and FIR spectroradiometers. A number of different band-pass filters may be arranged in a *filter wheel* (Figure 9-2a), which places a specified filter in the optical path by rotating to a specified position. The band-pass characteristics of spectrophotometer filters can be established by a composite glass absorption filter or by an interference filter. *Composite glass absorption filters* consist of one or more high-pass filters and one or more low-pass filters, with some overlap in their respective cutoff regions. The passband lies in that overlap region. The *band-pass* characteristic of a filter is usually given by the range of wavelengths at one-half the peak transmittance (FWHM = full width at half magnitude). Other essential filter characteristics are the transmittance at the nominal wavelength (*peak transmittance*), typically stated in percent of the transmittance that would exist without the filter (or with a clear glass plate) and the wavelength at which the peak is observed (*nominal wavelength*). For most composite glass absorption filters, the FWHM is around 25 nm and the peak transmittance is less than 20%. Single high-pass or low-pass sharp-cutoff filters are sometimes used in the optical path of other types of filters or other types of monochromators, as blocking filters.

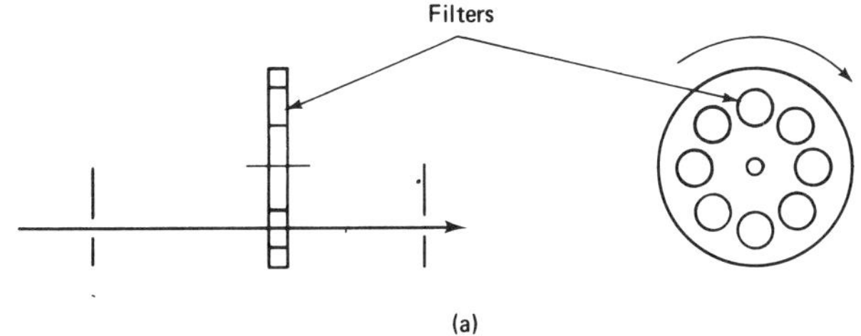

(a)

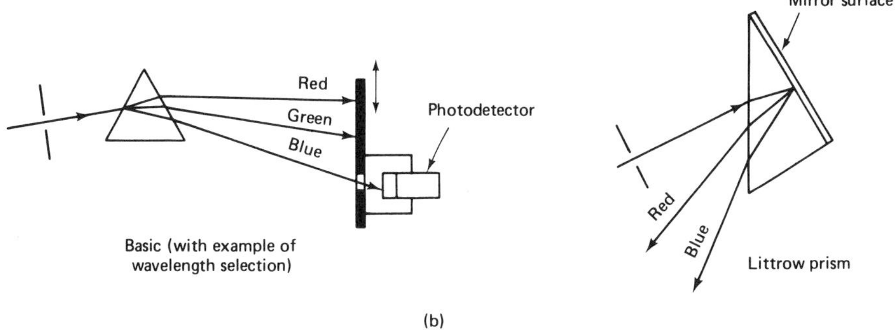

(b)

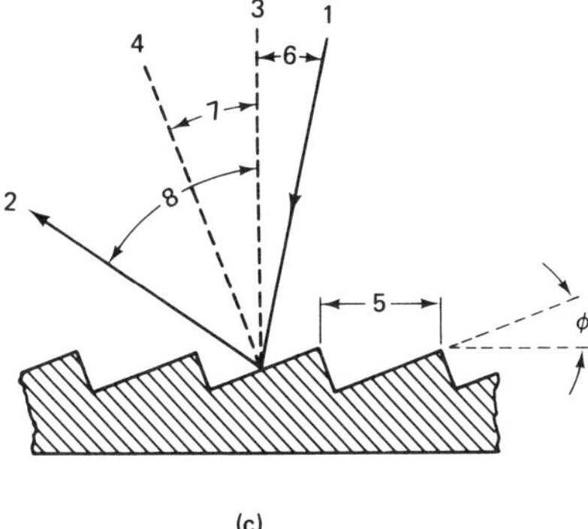

(c)

Figure 9-2. Monochromator types: (a) filter wheel; (b) prism; (c) diffraction grating: 1, incident beam; 2, diffracted beam; 3, normal to grating; 4, normal to groove face; 5, groove spacing (d); 6, angle of incidence; 7, blaze angle (ϕ); 8, angle of reflection.

Interference filters employ the principles of optical interference to reject radiation at wavelengths outside a pass band by selective reflection. This reflection occurs within a thin film (½ wavelength thick) of transparent metal and dielectric layers between two glass plates. Such filters have a FWHM of around 10 nm and a peak transmittance of around 50%. *Multilayer interference filters,* essentially several reflecting layers separated by glass plates, can have a FWHM of about 3 nm and a peak transmittance around 65%. A *filter wedge* is a continuously variable interference filter, with a wedgelike layer of dielectric between semitransparent metallic layers covered by glass plates. The filter wedge is moved, if rectangular, over the exit slit to select the desired nominal wavelengths; if circular, it is rotated over the exit slit to accomplish the wavelength selection.

The *prism* is probably the best known type of monochromator (see Figure 9-2b). The basic prism was used in some of the earliest instruments; wavelengths can be selected by moving the exit slit or by rotating the prism. Light dispersion is caused by the different angles of refraction for different wavelengths. There are several different prism designs that are used to varying extents in spectrophotometers. The *Littrow* prism (prism in a Littrow mount) has either a mirror placed behind it or has its rear surface mirrored (e.g., by deposition of an aluminum film) so that refraction occurs twice since the once-refracted waves are refracted again after reflection from the mirror. The *Pellin-Broca* prism, of irregular-trapezoid shape, produces a spectrum that is always at 90° to the incident beam. Two half-prisms can be cemented together back to back, or can be mounted facing each other at a specific angle with the optical path.

The *grating* (*diffraction grating*) is now the most frequently used dispersion device in monochromators. It consists of a flat plate or disk into which a large number of equidistant parallel grooves have been formed in its highly polished surface. Between 500 and 2500 grooves per millimeter are ruled (*blazed*) on a grating. The characteristics of a grating are illustrated in Figure 9-2c. The incident light beam is diffracted by the groove. The diffraction causes a spreading of light from the groove over a range of angles. At some of these angles, light at a specific wavelength is much more intense than the light at other angles due to constructive interference (reinforcement). Actually, several orders of spectra are produced by diffraction from the groove. For a given angle of incidence, the first-order diffraction spectrum will contain a set of wavelengths increasing with increasing angle of diffraction, the second order will contain a set of wavelengths one-half as long as those in the first-order spectrum, and the third-order spectrum will contain a set of wavelengths each of which, at the same angle of diffraction, will be one-thing as long as its first-order counterpart, and so on.

The characteristics most frequently specified for grating monochromators are the number of grooves per millimeter, the focal length, the overall range of wavelengths, the resolution obtainable in the first-order spectrum, and, for the grating itself, the *blaze wavelength,* the wavelength (first-order) for which the angle of diffraction and the angle of reflection are the same, and assuming that the angle of incidence (see Figure 9-2c) is the same as the angle of diffraction. The blaze wavelength is then dependent on the groove spacing and the sine of the blaze angle:

$$n\lambda_\phi = 2d \sin \phi$$

where n = order number
λ_ϕ = blaze wavelength
d = groove spacing
ϕ = blaze angle

The operation of a grating in a monochromator is illustrated in Figure 9-3. The grating is rotated by a drive and the band of wavelengths of interest is scanned by sweeping the diffracted light over a very narrow exit slit so that the photodetector senses (nominally) one wavelength at a time. What wavelength it senses is given by the grating characteristics, which are known, and the angular displacement of the grating, which is either measured (e.g., by an

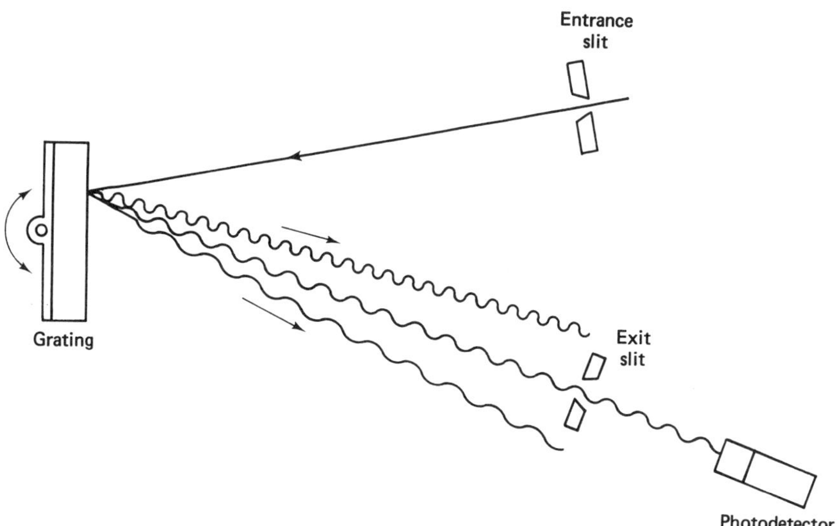

Figure 9-3. Basic grating monochromator operation.

angular-displacement transducer) or whose rate with respect to time is so accurately established that it can be correlated with time. Diffraction gratings can provide resolution better than 0.01 nm and are generally usable for wavelengths between 100 nm and 50 μm, with some special UV designs extending the range down to about 30 nm and (with the grating used in a grazing-incidence mode) as low as 1 nm. Developments are continuing to improve grating performance.

Various optical mounts are used in grating monochromators. Figure 9-4 illustrates two frequently used arrangements; modifications of both of these mounts exist.

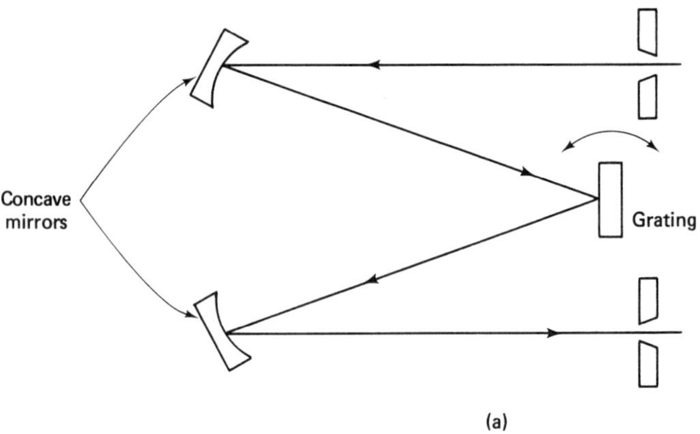

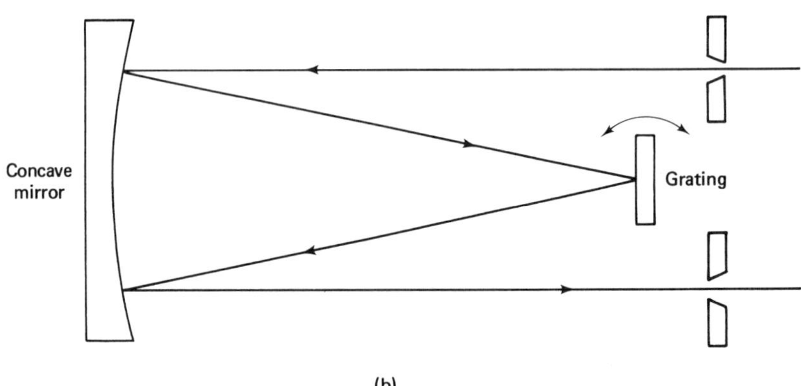

Figure 9-4. Typical grating monochromator mounts: (a) Czerny-Turner mount; (b) Ebert-Fastie mount.

9.2.3 Other Optical Elements

Besides the monochromator section with its entrance and exit slits, other optical elements include windows, lenses, flat and parabolic mirrors, and beam splitters. It is important that all optical elements have a high transmittance. Mirrors are coated on their front surface, typically by aluminizing; a coating may be applied over the aluminized surface to improve reflection or provide UV enhancement. Beam splitters are usually dichroic mirrors of semitransparent material which have one surface reflection-coated; they are mounted at 45° to the incident beam and will transmit certain wavelengths while reflecting other wavelengths at an angle normal to the beam.

9.2.4 Photodetectors

Photodetectors can be classified into two major categories: *Photon detectors* depend on effects produced when quanta of incident radiation (photons) react with electrons in a sensor material. *Thermal detectors* respond to total incident radiant energy and they are used primarily for IR sensing. Photon detectors employ photovoltaic, photoconductive, photoconductive-junction, photoemissive, or photoelectromagnetic transduction. Thermal detectors use thermoelectric, bolometric, or pyroelectric transduction methods. These transduction methods are illustrated in Figures 9-5 and 9-6 (where shown, e = electrons, $h\nu$ = photons or radiant energy). Finally, a third but minor category comprises pressure-actuated detectors.

9.2.4.1 Photovoltaic detectors. Photovoltaic detectors are self-generating; that is, they require no external excitation power. Their output signal (voltage, current, or power) is a function of the illumination of a junction between two dissimilar materials (Figure 9-5a). Several types of pairs of materials exhibit the photovoltaic effect, such as iron/selenium (as in the classic selenium cell, the earliest photocell) and copper/copper oxide. Modern photovoltaic detectors are made from semiconductor materials. Most of these are homojunction detectors (as explained in Figure 9-5a). Materials such as silicon, germanium, indium arsenide, or indium antimonide are made dissimilar by introducing different types of impurities (dopants) into opposite ends of the material. This makes one portion of the material *p-type* and the other *n-type,* and the junction between them, the *p-n junction,* then becomes the potential barrier. The *p on n* types are most common, with the *p layer* exposed to the incoming photons. An example of a heterojunction detector is the *n on p* type; it uses *n-type* glass deposited on *p-type* single-crystal silicon.

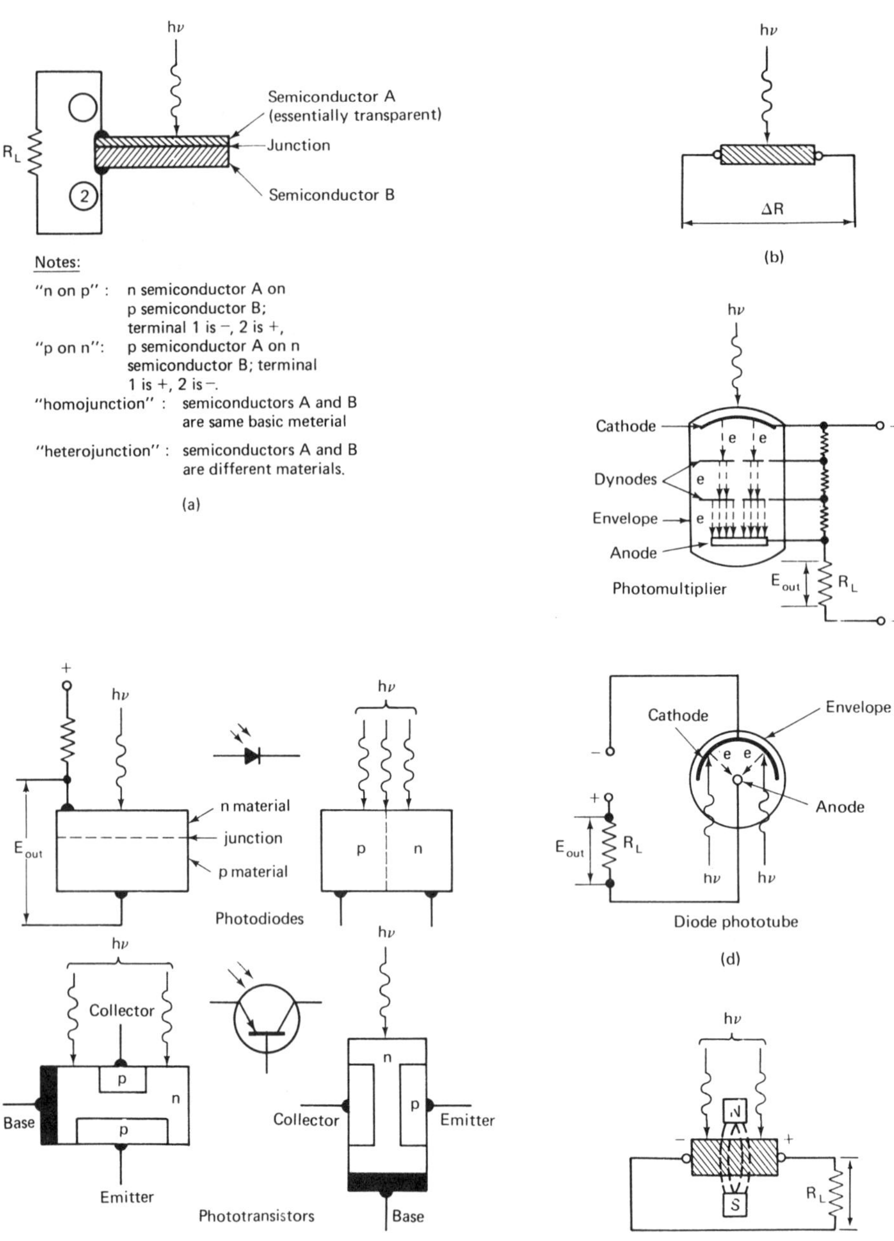

Figure 9-5. Basic light-sensing methods (photon detection): (a) photovoltaic; (b) photoconductive; (c) photoconductive-junction; (d) photoemissive; (e) photo-electromagnetic.

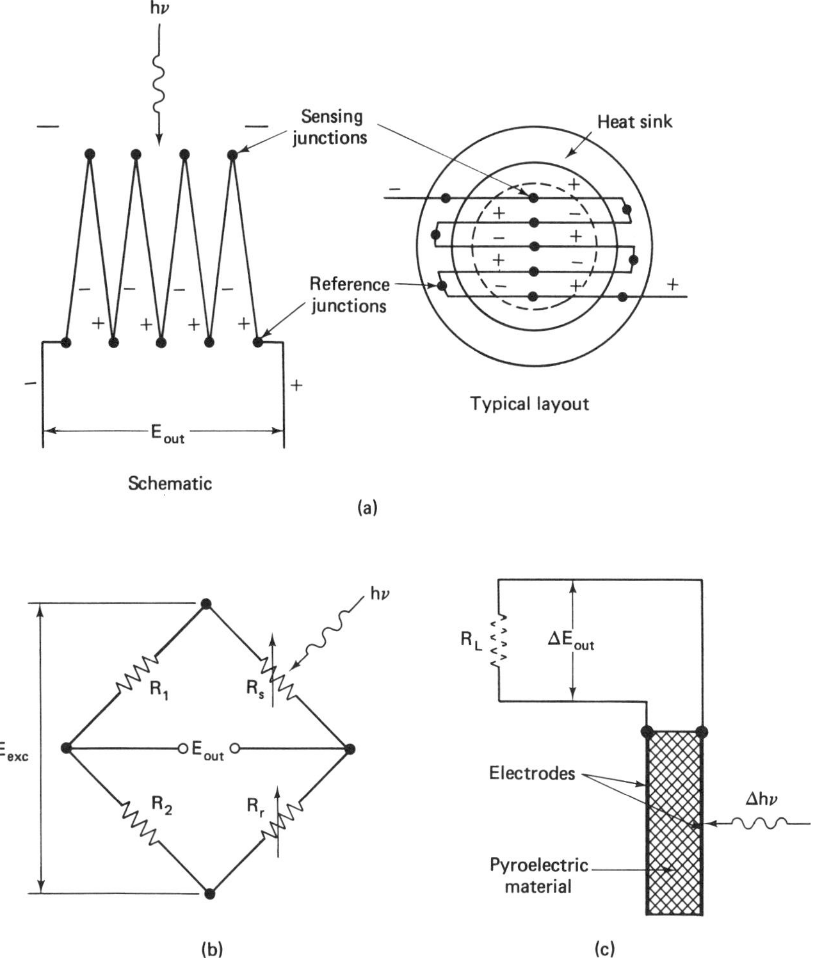

Figure 9-6. Basic light-sensing methods (thermal detection): (a) thermoelectric (thermopile); (b) bolometric; (c) pyroelectric.

9.2.4.2 Photoconductive detectors. Photoconductive detectors (Figure 9-5b) are made of semiconducting materials which reduce their resistance in response to incident illumination. The material is contained between the two conductive electrodes to which connecting leads can be attached. The change in conductance results from a change in the number of charge carriers created by the absorption of incident photon energy. Polycrystalline films, such as lead sulfide and lead selenide, as well as bulk single-crystal materials (doped germanium and silicon) are used as photoconductors. The cadmium sulfide (CdS) photoconductive detector has often been used in cameras, for automatic exposure control. Some materials (for example, mercury cadmium telluride) are used in the photoconductive as well as the photovoltaic mode.

9.2.4.3 Photoconductive-junction detectors. In the version of photoconductive detection called photoconductive-junction detection, the resistance across a junction between p and n semiconductor material changes as a function of incident light. The junction photocurrent (the leakage current varied by electron-hole pairs generated at or near the *depletion region*) increases with increasing incident photon flux. This principle is employed in *photodiodes* as well as *phototransistors* (Figure 9-5c).

9.2.4.4 Photoemissive detectors. Photoemissive detectors emit electrons from a cathode when photons impinge on it (Figure 9-5d). The electrons are ejected from the cathode surface when the energy of the radiation quanta is greater than the work function of the cathode material. This effect is used in the diode phototube (either evacuated or gas filled) as well as in the photomultiplier tube. In the diode phototube some of the electrons are collected by an anode which is at a positive potential with respect to the cathode. This causes current flow which can be used to produce an output voltage across a load resistor in series with the anode. In the photomultiplier tube additional electrodes (dynodes), at sequentially higher positive potential, are located between the cathode and the anode so as to amplify the electron current by means of secondary emission from the dynodes.

9.2.4.5 Photoelectromagnetic detectors. Photoelectromagnetic detection is a specialized detection method that is effected in a semiconductor that is acted upon by an external magnetic field (see Figure 9-5e). When photons are absorbed near the front surface of the semiconductor, the resulting excess of carriers at that surface and their relative absence at the rear surface cause a diffusion of carriers toward the rear. The force due to the application of a transverse magnetic field then deflects the holes toward one end of the semiconductor and the electrons toward the other end, thus causing an emf to be developed between the two end terminals.

9.2.4.6 Thermoelectric detectors. The thermoelectric (Seebeck) effect is employed in miniature multijunction thermocouple devices called *thermopiles*. The thermopiles, which consist of a number of thermocouples connected in series, produce an output voltage when the temperature of their sensing junctions is higher than the temperature of their reference junctions. Thermopiles used for radiant-flux sensing have their reference junctions in thermal contact with a heat sink, whereas their sensing junctions are blackened (to absorb heat radiation) and thermally isolated from the heat sink (Figure 9-6a). The reference (cold) junctions remain relatively stable in temperature; they are also shielded so as not to receive any of the incident radiant flux. The sensing junctions are heated by the radiant flux. If the heat-sink temperature is known, the temperature difference between the reference and sensing (hot) junctions, as indicated by the thermopile output voltage, is a measure of the incident radiant flux.

9.2.4.7 Bolometric detection. Bolometers used for radiant-flux sensing generally consist of a pair of matched thermistors (or other temperature-sensitive resistive material such as silicon, indium antimonide, or single-crystal gallium-doped germanium) connected in a half-bridge or full-bridge circuit. One of the thermistors is blackened and mounted in such a manner that it senses radiant flux, while the second one is isolated from radiant flux and responds only to the temperature of the heat sink. A typical bolometer circuit is shown in Figure 9-6b, where R_s is the radiation-sensing thermistor, R_r is the reference thermistor shielded from radiation, and R_1 and R_2 are a pair of matched, stable resistors, usually located some distance away from where thermal radiation is being sensed. When an excitation voltage E_{exc} is applied to the circuit, the output voltage will be proportional to the difference in resistance between R_s and R_e, which, in turn, is a measure of incident radiant flux. If this were a half-bridge circuit, E_{exc}, R_1, and R_2 would be replaced by two excitation voltages and exactly the same amplitude but opposite polarity, inserted in the place of R_1 and R_2.

9.2.4.8 Pyroelectric detectors. Pyroelectric detectors are composed of a ferromagnetic crystal material (e.g., triglycene sulfate, $LiTaO_3$) between two electrodes. The crystal exhibits a spontaneous polarization (electric charge concentration) which is temperature dependent. Changes in incident radiant flux, absorbed by the crystal, cause a change in the crystal temperature, resulting in a change in the potential difference (voltage) across the electrodes. This voltage is later neutralized by current flow through the internal leakage resistance and the external load resistance. Figure 9-6c illustrates the pyroelectric sensing method. It should be noted that materials other than solid crystals, notably certain polymers, also exhibit the pyroelectric effect.

9.2.4.9 Pressure-actuated photoelectric detectors. Pressure-actuated photoelectric detection is a transduction method that is employed in the *Golay-type IR detector* (*Golay cell*). A gas of low thermal conductivity (e.g., xenon) is enclosed in a cylinder capped by a blackened membrane on one end and a mirror-coated diaphragm at the other end. IR radiation incident upon the blackened membrane causes the gas to expand and the mirror-surfaced diaphragm to deform. The diaphragm is positioned in the optical path between a light source and a light sensor in such a manner that deformation of the diaphragm causes the light reflected by the mirrored surface toward the light sensor to increase with increasing pressure. The output of the light sensor is then proportional to incident IR radiation.

A somewhat related device, in that it also uses a diaphragm but employs capacitive rather than photoelectric transduction, is the *Luft detector.* This detector is used primarily in some designs of dual-beam NDIR analyzers (see Section 7.10). The two IR beams, one through the sample cell, the other through the reference cells, are made to fall on opposite sides of a diaphragm. The diaphragm thus deflects proportionally to the difference in the two IR energies. This deflection causes a change in capacitance between the diaphragm and a static electrode. In a typical design, the diaphragm is in a sealed, gas-filled chamber provided with windows for the two beams. The difference in thermal energies causes a corresponding difference in pressure, resulting in diaphragm deflection. As is typical for most low-pressure diaphragm-type sensors (e.g., condenser microphones), this type of detector is sensitive to vibration as well as to variations in ambient temperature.

9.2.5 Spectrophotometer Instruments

Industrial spectrophotometers are generally absorption spectrometers which can be used for single-component or multi-component analyses. The basic elements of a spectrophotometer are a light source, a monochromator, and a photodetector. The monochromator can be a filter (interchangeable with filters for other wavelengths), a filter wheel with numerous filters, or a prism or grating. A typical light source is the tungsten-halogen lamp; however, other light sources (e.g., deuterium for UV) can be used when the spectrum of interest extends below or significantly above the visible-light range. At least one design, however, uses a *polychromator* instead of a monochromator; the entire spectrum is then seen by an array of a large number of detectors, with each detector positioned to respond to a narrow band of wavelength. Typical detectors include the photomultiplier tube (190–600 nm), and silicon (200–1000 nm), germanium (800–1600 nm), or lead sulfide (1000–3200 nm) detectors.

The sample can be placed into the optical path either after (*predispersive*) or before the monochromator, depending on instrument design and the type of analysis to be performed. For single-component analyses, the sample is usually placed between the monochromator and the detector, and the monochromator can be a single (but interchangeable) filter or a prism or grating movable to a predetermined position. For multi-component analyses the prism or grating operates in a scanning mode, moved by a motor with a cam mechanism.

Single-beam spectrophotometers have a single optical path between the source and the detector. *Double-beam* instruments incorporate a reference channel that is sample-indifferent. This is often accomplished by switching the beam between two filters, one especially selected to pass a wavelength band sensitive to the sample, the other, sample-indifferent. This system uses a single detector. Another way of accomplishing this is to have a second detector (or one or more elements of a detector array) that responds to a sample-indifferent wavelength. If a scanning monochromator is used, the sample-sensitive as well as the sample-indifferent wavelength are both known from the angular position of the monochromator. If a polychromator and multi-element detector array are used, the sample-sensitive and sample-indifferent wavelengths become obvious from the spectrum displayed. In double-beam spectrophotometers, it is usually the *ratio* of the sample-sensitive reading to the sample-indifferent reading that is displayed and recorded.

Modern spectrophotometers use a CRT or LCD display (additional hard-copy devices such as a strip-chart recorder are usually optional) as well as built-in programmable processors and interfaces to other computers.

A single-beam spectrophotometer for the primarily visible-light range (400–900 nm) is illustrated in Figure 9-7. The instrument uses a gas-filled tungsten lamp as source and a UV-enhanced silicon photodiode as detector. The appropriate wavelength is selected by manually adjusting the angular position of the Littrow prism, using a vernier knob. Wavelength accuracy and resolution are reported as ± 2 nm and ± 1 nm, respectively. The instrument incorporates a processor and memory that allows any of over 120 preprogrammed calibrations (conversion of absorbance data to concentration values) to be selected. The LCD readout can be programmed to display concentration, absorbance, or % transmittance. Applications are primarily in water and soils analysis, aquaculture (fish farms), and surface finishes.

The dual-beam spectrophotometer illustrated in Figure 9-8 employs a motor to drive a filter wheel containing two filters, 180° apart. One filter passes the measuring wavelength selected (at which the component of interest has a characteristic spectral peak); the other passes a reference wavelength (at which none of the sample components has an absorbance peak). The filter wheel also acts as chopper. The instrument operates in the near-IR band of 1.0 to 2.7 μm and uses a quartz iodide light source. The instrument's electronics converts the two pulses of optical energy seen by the detector into two electrical pulses (pulses of optical energy result from the use of a chopper); the circuitry further takes the log ratio of the measured-to-reference pulse amplitudes, so that the net output signal is linearized and specific to the

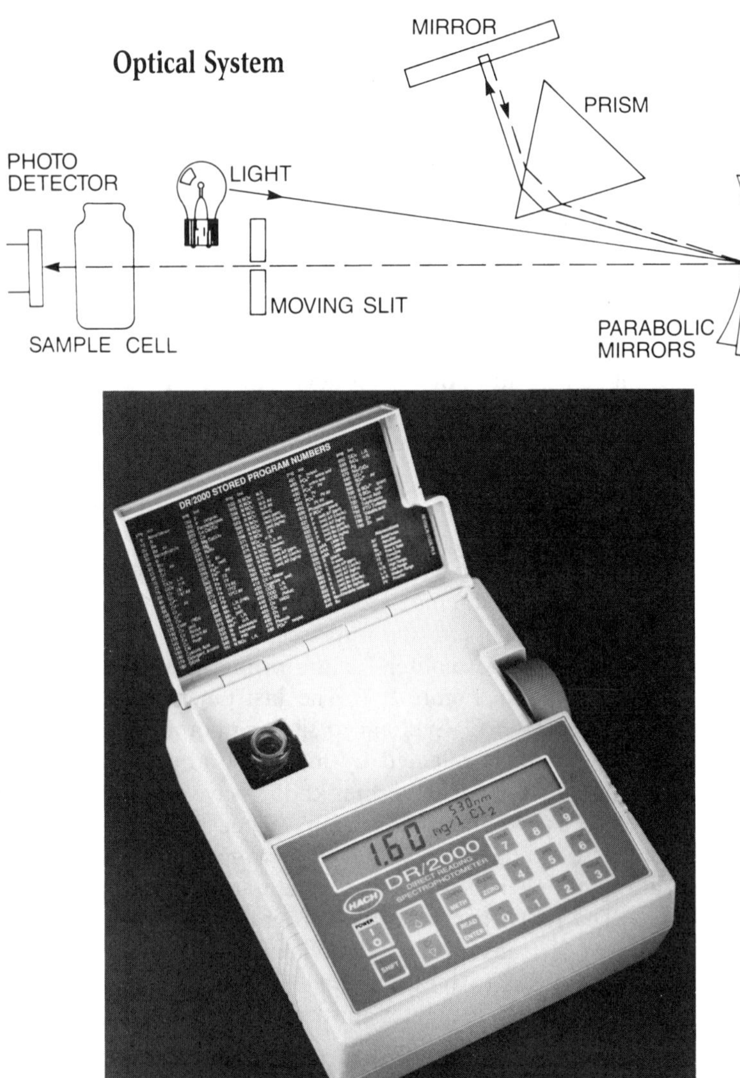

Figure 9-7. Single-beam visible-light spectrophotometer. (Courtesy of Hach Co.)

component of interest. The same type of optical system is also used in a UV/Visible spectrophotometer (210–1000 nm).

The use of fiber optics allows analyses to be performed on samples located remotely from the instrument, such as in hazardous or relatively

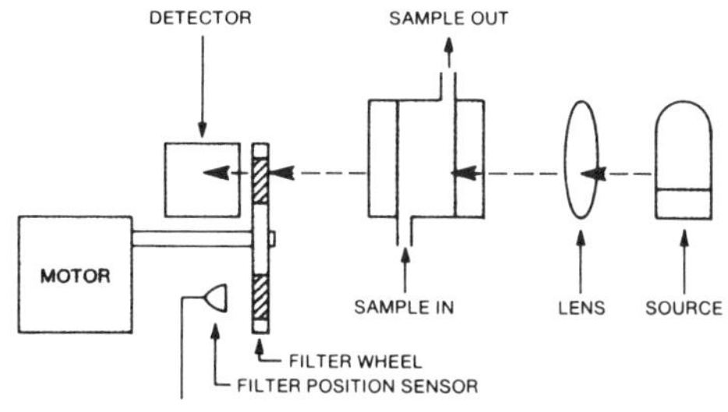

(a) Optical path

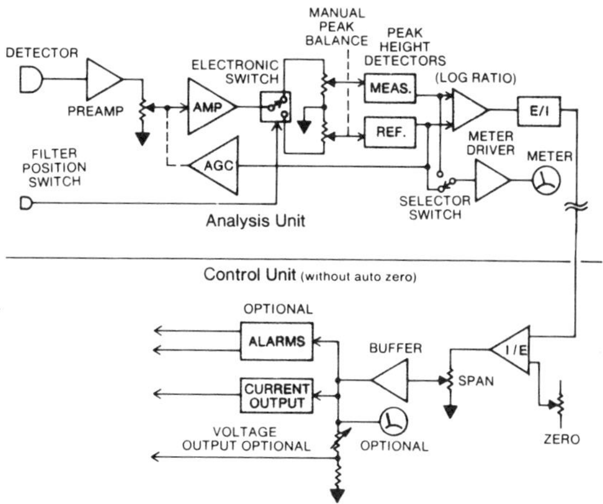

(b) Electronics block diagram

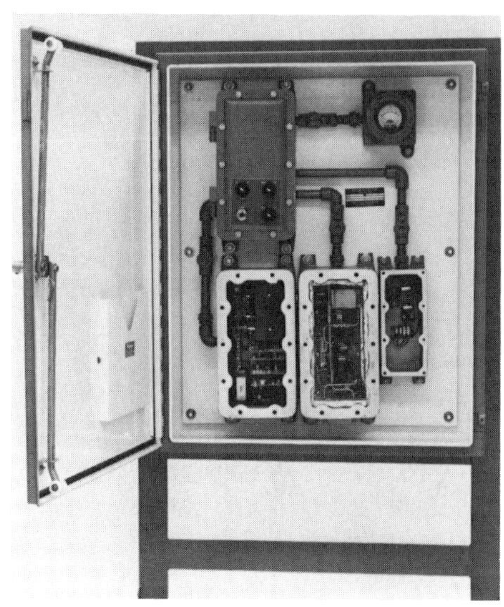

Figure 9-8. Dual-beam near-infrared spectrophotometer. (Courtesy of Teledyne Analytical Instruments.)

inaccessible areas. Distances between sensing probe and instrument can easily be up to 10 m and can be extended up to 500 m. Figure 9-9 shows the optical path of a scanning spectrophotometer using fiber optics. A tungsten-halogen lamp is used as the light source (a deuterium light source is optionally available for analyses in the UV range). The light beam is collimated unto the sending fiber. Multimode silica fibers or fiber bundles are employed in the sending as well as receiving cables. The sending and receiving cables are attached to a remotely located probe or set of two probes, depending on the application. The sample-modified light travels through the receiving cable and is introduced to the monochromator, a servo-driven scanning holographic grating. Two photodetectors with overlapping spectral response are used in most versions of this design, depending on the overall spectral range of the instrument. Detector No. 1 responds to longer wavelengths than detector No. 2. Depending on model number, wavelengths range from 250–600 nm (using only a photomultiplier tube as detector) or 250–800 nm (using a photomultiplier tube as detector No. 2 and a silicon photodetector as detector No. 1) to 1100–2400 nm (using a PbS detector as No. 1 and a Ge detector as No. 2). Gratings are available with 1200 lines/mm (for shorter wavelength ranges) or 800, 600 or 300 lines/mm (the latter for the longest wavelength ranges).

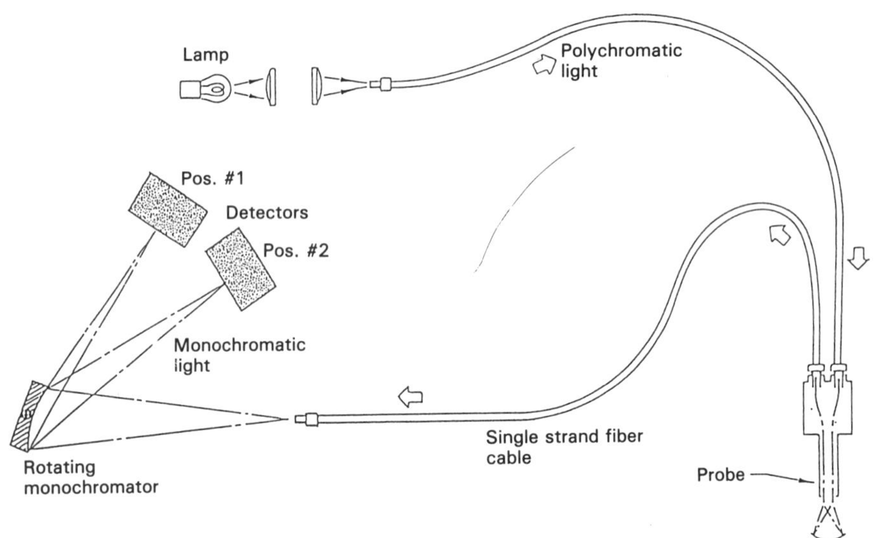

Figure 9-9. Fiber-optic spectrophotometer—optical path. (Courtesy of Guided Wave, Inc.)

Figure 9-10 illustrates some of the large variety of different probe designs available. In the transmission probe, the sending and receiving fiber optic terminations (with sapphire windows as collimators) face each other; the path length can range up to 50 mm. The wand probe, which exists in a variety of configurations, employs a removable mirrored tip for absorbance measurements. Without the tip, the probe can be used for measuring reflectance or fluorescence from the sample. The variable-path probe (the path length is user-adjustable) is typically used for transmission and absorbance measurements on translucent materials such as polymer films, glass sheets, and plant leaves.

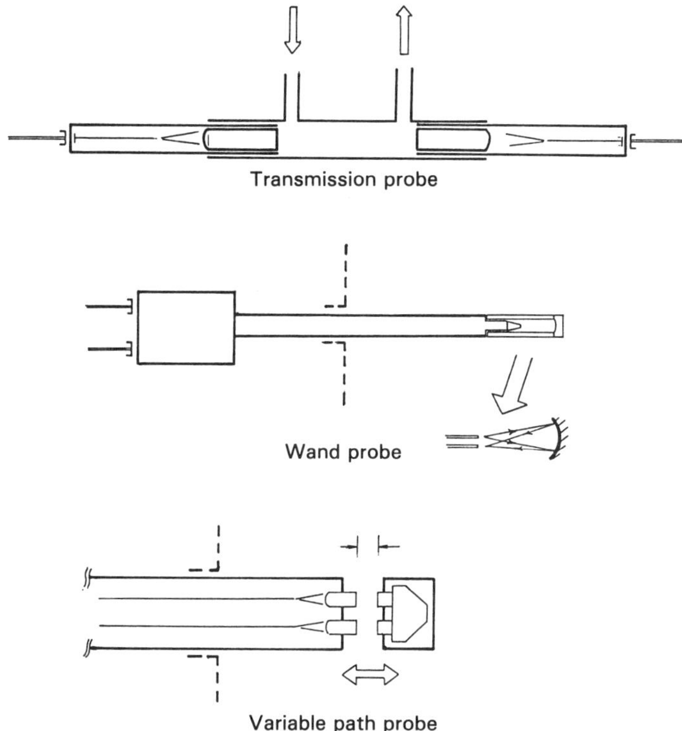

Figure 9-10. Fiber-optic spectrophotometer—typical remote-sensing probes (Courtesy of Guided Wave, Inc.)

Figure 9-11 shows a block diagram of the spectrophotometer system; a transmission probe is used as example. The electronics (computer) section provides signal processing and optics control. The cleanup portion smoothes the data, and the predicator performs predictions for up to ten constituents

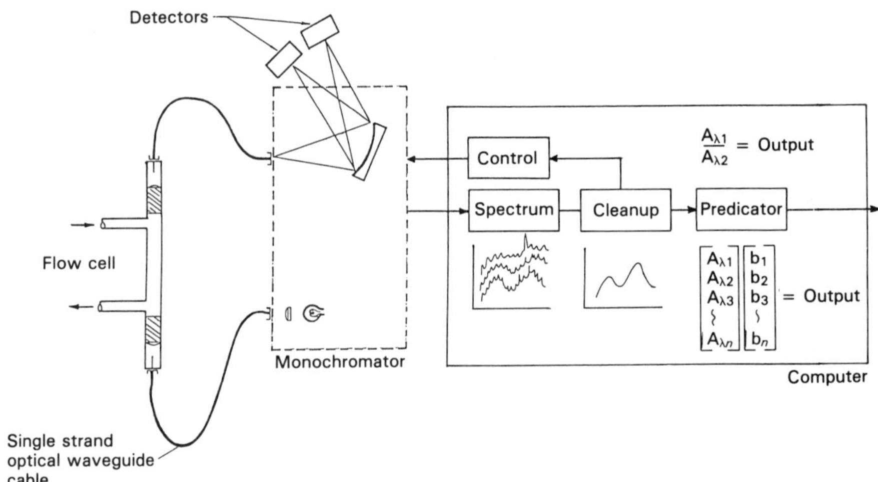

Figure 9-11 Fiber-optic spectrophotometer—instrument system block diagram. (Courtesy of Guided Wave, Inc.)

when the wavelengths of specific absorptions are known. Additional software is available for other functions. The software assumes that ancillary equipment is PC compatible and includes an EGA monitor. Examples of spectra obtained with this instrument are shown in Figure 9-12.

The spectrophotometer shown in Figure 9-13 is designed specifically for color measurements (see Section 7.2) and is therefore called a *spectrocolorimeter.* It differs from other designs discussed in this section primarily in that it has no moving parts (filter wheel, prism, or grating). Polychromatic light from a tungsten-halogen lamp (a UV-enhanced source is optionally available) enters an integrating sphere, which is internally Halon-coated for high reflectance. The instrument can operate in the transmittance or reflectance mode.

Light transmitted through the sample, or reflected from the sample, enters the polychromator chamber (see Figure 9-13b). The major elements in this figure are identified by numbers. Light enters the chamber through a lens system (10b) which collimates the light into a thin horizontal line which is incident upon the holographic diffraction grating (11). This grating disperses the light energy into its component wavelengths (375–750 nm) and directs them onto an array of silicon diode detectors (12). This array contains 76 elements (for an optionally available range extension to 1100 nm an array of 146 elements is used). Hence, the array offers a spectral resolution of 5 nm. The response time of each element is 16 ms, and the element outputs are scanned sequentially 228 times to allow signal averaging. For calibration pur-

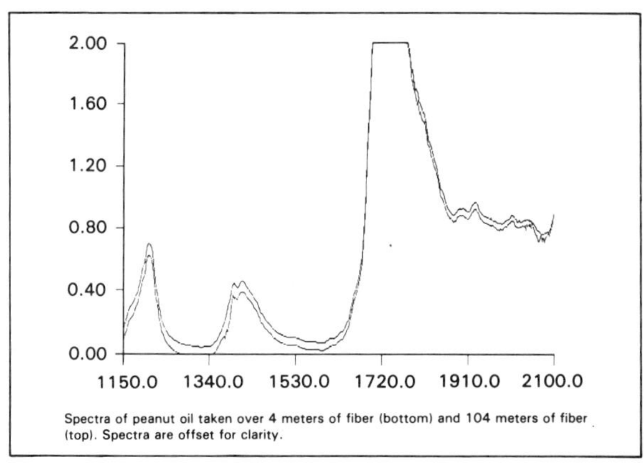

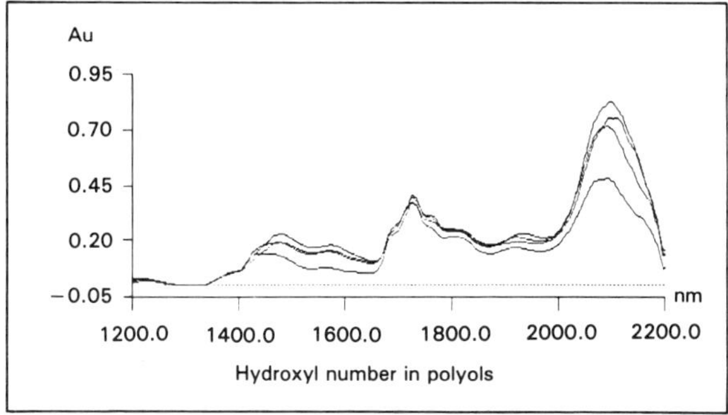

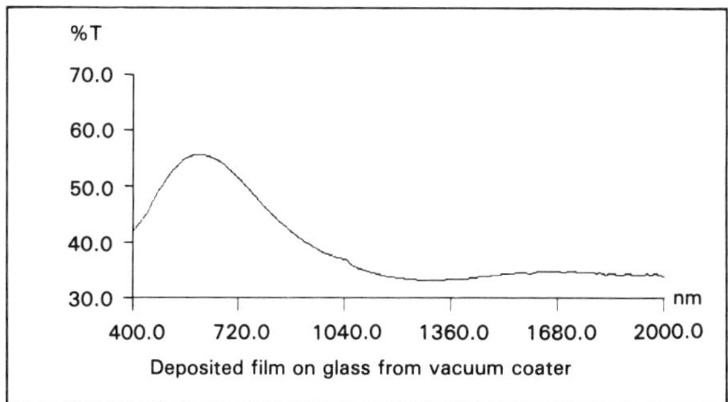

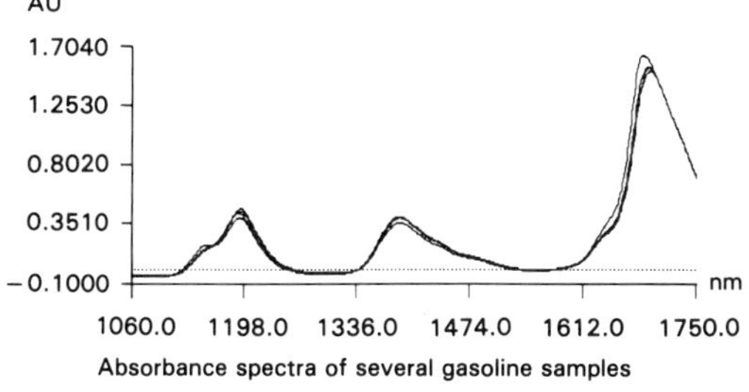

Figure 9-12. Typical spectra obtained by fiber-optic spectrophotometer (Courtesy of Guided Wave, Inc.)

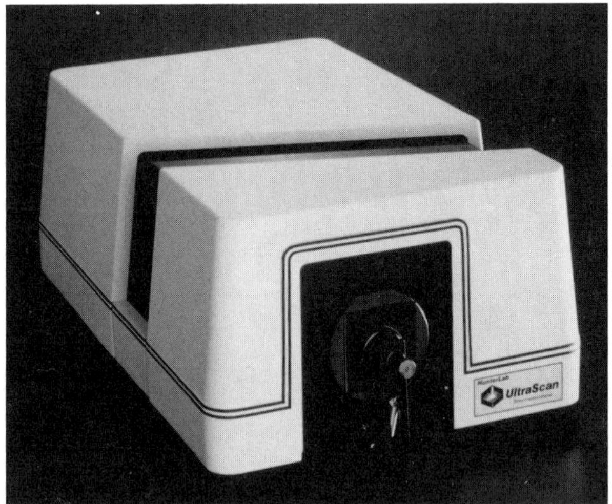

Figure 9-13. Spectrophotometer with fixed-grating polychromator and multidetector array: (a) instrument assembly; (b) polychromator chamber. (Courtesy of Hunter Associates Laboratory, Inc.)

poses, the instrument can be switched into its ACLC (automatic closed-loop calibration) mode, in which fiber optics are used to direct the light from the source directly to the polychromator. The instrument electronics include a serial RS-423 communications port, which is compatible with an RS-232 bus that can be linked to a remotely-located computer.

Infrared spectrometers related to spectrophotometers should properly be called *infrared spectroradiometers*. Spectrophometers whose spectral range extends up to 3000 nm (3 μm) have already been discussed. They differ from shorter-wavelength designs primarily in the choice of their photodetector. Infrared spectroradiometers for the middle-IR (3–15 μm) and far-IR (15–800 μm) regions are rarely used industrially but have been designed for research purposes, especially in the area of remote sensing; however, it should be noted that researchers are increasingly switching to interferometer-type designs (see Section 9.3). The similarity of middle- and far-IR spectroradiometers to conventional spectrophotometers extends to instruments that operate well into the middle-IR since quantum (photon) detectors, such as HgCdTe (when cooled), are usable to nearly the long-wavelengths limit of this region. However, monochromator gratings may not cover a spectral range extending to 15 μm. Hence, two or more grating-filter sets, designed for ease of interchangeability, are employed.

Far-IR spectroradiometers are still relatively similar to spectrophotometers up to about 30 μm. A combination of a grating and filters, mostly multilayer interference filters, is used as monochromator; however, several grating/filter sets, designed to be interchangeable, must be used if a wide range of wavelengths, extending to about 30 μm, is to be scanned. The detector is almost invariably of the thermal type (i.e., thermopile, bolometer, or pyroelectric); the Golay detector is still used in some instruments. Extrinsic silicon detectors (gold- or bismuth-doped) have been developed that are usable up to 30 μm if they are maintained at a temperature of 2 to 4 K. Successful operation of such detectors at long wavelengths requires not only cooling of the detector but also of the filters and, to a somewhat lesser extent, of the optics. As the wavelength to be analyzed increases above 30 μm, the photon energy decreases to such low levels that transmission losses (even those of good filters), associated with narrow-band-pass characteristics, can be tolerated less and less. The filters used will still tend to be multilayer interference filters, but their bandpass will get broader. Above about 50 μm gratings lose their usability and spectral selection is performed by a set of filters. These can be mounted in a filter wheel and used in conjunction with a thermal detector; however, better performance seems to be attainable when several detectors, each provided with its own filter, or set of filters, are used.

9.3 INTERFEROMETER SPECTROMETERS

The use of interference techniques has been applied successfully to spectrometers, and it has been found particularly useful in high-resolution IR spectrometers for wavelengths to about 50 μm. The *Fabry–Perot interferometer* has been used in spectroscopic applications, including in determinations of hyperfine spectra. It operates on the principle of multiple reflection. It employs two glass plates, spaced some distance apart, each having a thin coating (silver or aluminum) on the surfaces facing each other. The two reflecting surfaces must be exactly parallel to each other. The incoming wave is multiply reflected between the two plates and ultimately transmitted. The spacing between the two plates may be varied (if this is done without affecting parallelism), for example, by spacers, whose length can be between 1 and 200 mm.

The *Michelson interferometer* is now employed in a number of IR spectrometers. An example of a Michelson interferometer is illustrated schematically in Figure 9-14. Such instruments are used in the process industries and in quality-control and research laboratories, in a variety of applications, as well as in atmospheric, planetary, and space research. As shown in the illustration, the beam of energy to be analyzed is divided by a beam splitter between a fixed mirror and a moving mirror. The beam splitter is made of IR-transparent material (e.g., KBr or CsI), with a semireflecting coating, below which is attached another slice of the same IR-transparent material to act as compensator, to equalize the optical paths in the two branches. The compensator is

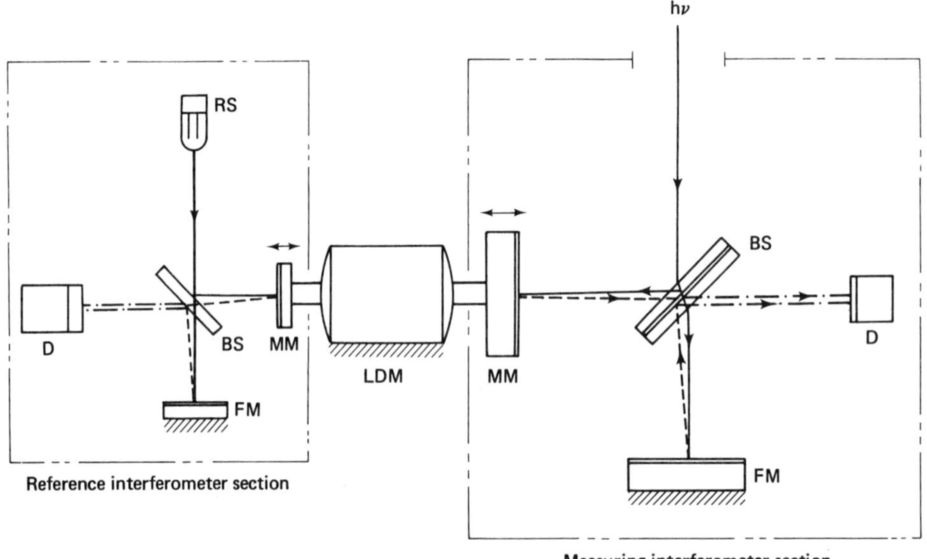

Figure 9-14. Michelson interferometer spectrometer: $h\nu$, incoming radiation to be analyzed; BS, beam splitter; MM, moving mirror (Michelson mirror); FM, fixed mirror; D, detector; RS, reference (light) source; LDM, linear-drive motor (Michelson motor).

constructed integrally with the beam splitter per se. Each mirror reflects its portion of the beam back to the beam splitter, where the reflected beams combine and the combination product is then sensed by the detector. When the two reflected beams are exactly equal in length, they will constructively interfere at the beam splitter and the detector will sense the reinforced light (radiation). When the Michelson mirror moves ¼ wavelength from this position, the two reflected beams will interfere destructively, because they are 180° out of phase, and the radiation sensed by the detector will be minimal. As the mirror moves another ¼ wavelength, the interference will again be constructive. Hence, the detector will see a brightness peak every ½ wavelength as the mirror moves. The output of the detector, when the incoming radiation is monochromatic at a certain wavelength, will be at a frequency equal to twice the velocity of the moving mirror divided by the wavelength. The variation of signal with mirror displacement follows a cosine relationship. When the incoming radiation is polychromatic, the output signal will be an interferogram (see Figure 9-15), the sum of all the cosine waves, the Fourier transform of the spectrum. Hence, the term *Fourier transform spectroscopy* has been applied to the technique employed by this type of spectrometer. An inverse transformation of the interferogram is used to reconstruct the spectrum; this is usually accomplished by means of a digital computer and an analog conversion device.

The two parameters critical to determining the resolution of an interferometer spectrometer are the length of the optical path between the

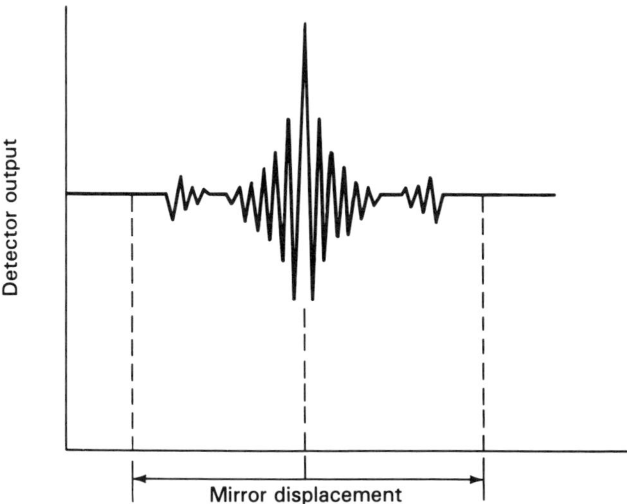

Figure 9-15. Typical interferogram.

beam splitter and the fixed mirror and of the optical path between the beam splitter and the moving mirror. The difference between these two path lengths is the *optical path difference* (*OPD*) or *optical retardation*. The resolution of the instrument, expressed in wave numbers, is the reciprocal of the OPD, expressed in centimeters. It is important that the mirror displacement is very accurately known during the operation of the instrument. The motion is usually generated by electronics that make the motion a precise function of time; the interferogram can then have time as the motion-related coordinate. Some such instruments employ a reference interferometer (see Figure 9-14) to produce a signal which is fed to the mirror-drive electronics to time-lock the mirror motion. The output of the reference interferometer detector can additionally be monitored to provide status information.

Interferometer spectrometers operate almost exclusively in the IR portion of the spectrum. Therefore, they are now commonly referred to as *Fourier Transform IR* (*FTIR*) analyzers. Industrial FTIR analyzers normally have a wavelength band significantly less than 50 μm, which simplifies the choice of optical materials; the beam splitter is then made of a less delicate material than KBr or CsI. A typical design contains a polychromatic IR source and places the sample just ahead of the detector. Gas analysis is generally the main function of such instruments. A related device is the *Hadamard-transform spectrometer,* which is based on dispersive rather than interferometric optics and operates on the principle of multislit optical encoding at the spectrometer exit focal plane.

9.4 RAMAN AND LASER SPECTROMETERS

When molecules of a gaseous, liquid, or solid compound are illuminated by a light source, they will scatter this light; most of the scattering products will be at the same wavelength (frequency) as the incident light; however, a small fraction of the scattering products will have undergone a wavelength change (frequency change) in the scattering process (*Raman effect,* named after its discoverer, C. V. Raman, in India, in 1928). The Raman-scattered light has no phase relationship with the incident light. The Raman spectrum, for any one molecule, will consist of several lines that are shifted in frequency by varying amounts from the incident-light frequency. This pattern is symmetric about the exciting line; however, the lines representing frequencies lower than the exciting line (*Stokes lines*) are always more intense than the corresponding lines on the high-frequency side (*anti-Stokes lines*). Since Raman spectra are based on frequency differences from the exciting frequency, they are usually shown with the abscissa graduated in wavenumber differences (Δ cm^{-1}), with zero indicating the excitation frequency; and, since only the

Stokes lines are displayed, zero is usually at the right end of the abscissa. Raman lines also show polarization to various degrees, depending on the origin of the line as well as on the optical geometry of the instrument. The excitation frequency must be chosen so that no absorption occurs in the compound analyzed. Raman spectrometers are very useful in analyses of molecular structure and behavior and for quantitative analyses of major components in complex mixtures.

The major elements of a laser spectrometer are a monochromatic light source, a scatter-collection mirror placed normal to the incident beam to collect the scattering products from the sample illuminated by the beam, collection optics, a scanning monochromator to enable analysis of the various frequencies (wavelengths) of the Raman-scattered light, and a photodetector.

Since a high-intensity light source is required to obtain measurable Raman scattering, a laser is used for this purpose in all modern instruments, replacing the formerly used mercury-arc lamp combined with elaborate filters. The laser wavelengths are usually in the visible-light range, and the instrument optics are of the types used in visible-light spectrophotometers. The laser is chosen for its wavelength and avalable beam intensity and the nature of the analysis. Commonly used lasers are He–NE (632.8 nm), argon (488.0 and 514.5 nm), and krypton (568.2 nm) (see Table 9-1). The laser beam is focused at the sample, using appropriate collimating optics that may include a multipass mirror behind the sample. The direction of observation is normal to the laser beam and suitable collection optics are used to focus the scattering products toward the monochromator entrance slit. The monochromator may be a double-grating or triple-grating type. Multiple monochromators help reduce stray light originating from the high-intensity excitation and the very intense Rayleigh scattering products at the excitation wavelength. A photomultiplier or other photon detector is used to detect the light from the monochromator exit slit.

TABLE 9-1. Typical Laser Wavelengths

Laser Type	Wavelength (nm)
Nd:glass (fourth harmonic)	266.0
He–Cd	325.0
Nitrogen	337.1
He–Cd	441.6
Argon ion	488.0
Argon ion	514.5
Nd:YAG (second harmonic)	532.0
He–Ne	632.8
Ruby	694.3
GaAs	905.0
Nd:YAG, Nd:glass	1064.0

9.5 ATOMIC ABSORPTION AND FLAME EMISSION SPECTROMETERS

This category of spectrometers employs either a spectrophotometer or a monochromatic photometer for analysis (i.e., radiation intensity at certain wavelengths).

In *flame emission spectrometers,* the elements used are the same as in the flame photometer (see Section 7.9) except that a monochromator is added ahead of the photodetector so that the spectrum can be analyzed for a number of different elements and molecules. The sample is vaporized and burned in the flame, and the monochromator in conjunction with the photodetector permits identification of constituents and their abundance by examination of peak wavelength and peak area.

In *atomic absorption spectrometers,* light from a hollow-cathode lamp, whose cathode contains the element of interest, is passed through a flame in which the vaporized sample burns. The relative abundance of the element of interest in the sample will be indicated by the degree of absorption, in the flame, of a specific spectral line. The absorption line is selected by appropriate adjustment of the monochromator (if it is continuously variable) or by inserting a narrow-band-pass filter. Among the many applications of such instruments is the detection of mercury in the atmosphere and in food products, in amounts of 1 ng or less; this is accomplished by a double-beam atomic absorption spectrometer; the reference beam does not pass through the sample (through the flame) and absorption at a characteristic mercury line is determined by a ratio method. There are, of course, numerous other applications of atomic absorption spectrometers.

In the less frequently used *atomic fluorescence spectrometers,* a light source irradiates the flame from an angle normal to the spectrometer optical axis and energy (first absorbed, then reemitted as fluorescence) is analyzed by the monochromator-detector system.

9.6 X-RAY FLUORESCENCE SPECTROMETERS

X-ray fluorescence (*XRF*) spectrometry is used increasingly for the analysis of constituent elements in materials including air-pollution products (on filter paper), blood, mineral ore, alloys, clay, and ceramic glazes. The spectrometer always includes a source, X-ray optics, a pulse-height analyzer and/or monochromator, and a detector.

The source is usually an X-ray lamp, which may incorporate two selectable targets (e.g., molybdenum and tungsten), with provisions for accurate adjustment of voltage and current. Alternatively, radioactive-isotope sources (e.g., ^{55}Fe, ^{109}Cd) can be used; these sources provide low-energy X rays or gamma rays to the sample; they may be annular in shape so that they provide a larger radiation-emitting surface to the sample. The source excites X-ray fluorescence in many elements; the mechanism involves expulsion of an electron from the atom's inner shell and emission of secondary X rays upon the return of the electron to its normal state. The energy (or frequency or wavelength) of the secondary X rays characterizes each element (each atom) in the sample; however, this method is best applied to the detection of elements having an atomic number of 9 (fluorine) or higher.

For many analyses it is sufficient to use an SCA or MCA (see Section 9.7) to obtain a usable spectrum display. In other cases, however, it is necessary to use a monochromator to obtain sufficiently fine spectral resolution (an SCA or MCA is then often used additionally). A flat or curved diffraction crystal usually provides the monochromatization, together with entrance and exit slits. In nondispersive XRF spectrometers, the crystal is stationary and acts mainly as a selectable-wavelength filter. Scanning crystals are used in dispersive instruments. More than one crystal, often interchanged by mechanical flipping, may be required. Typical crystal materials (selected for their spectral characteristics) include lithium fluoride, potassium iodide, ammonium dihydrogen phosphate (ADP), and other phosphates and tartrates.

The detector for the secondary X rays is now typically of the semiconductor type, such as Si(Li) and Ge(Li), which tend to operate best when cryogenically cooled, and HgI_2, which operates adequately at room temperature.

Figure 9-16 illustrates an XRF spectrometer designed for alloy analysis. It covers the identification of 21 elements that are components of virtually all metal alloys. The instrument system consists of a probe connected to a data processing unit by a cable. The probe uses a dual isotope source (^{55}Fe and ^{109}Cd) and a mercuric iodide (HgI_2) detector; the aperture window is made of 25 μm thick polypropylene. The probe, usually hand-held, has a mass of 1.1 kg. The system is portable and battery-powered but contains an ac charger for either ac operation or battery recharge. The probe uses a 3.6 V lithium battery, user replaceable (one year nominal life). The data processing unit, microprocessor-controlled through a keyboard interface and a menu-style instruction format, allows three different operating modes, depending on sample size and configuration, as well as a reference mode used for calibration. Its memory contains an alloy library with 200 prestored entries. The display is selectable so that it shows the alloy name, or the component elements, each identified as percent by weight, or a signature match to a known sample. The normal time to acquire an analysis is about 25 seconds.

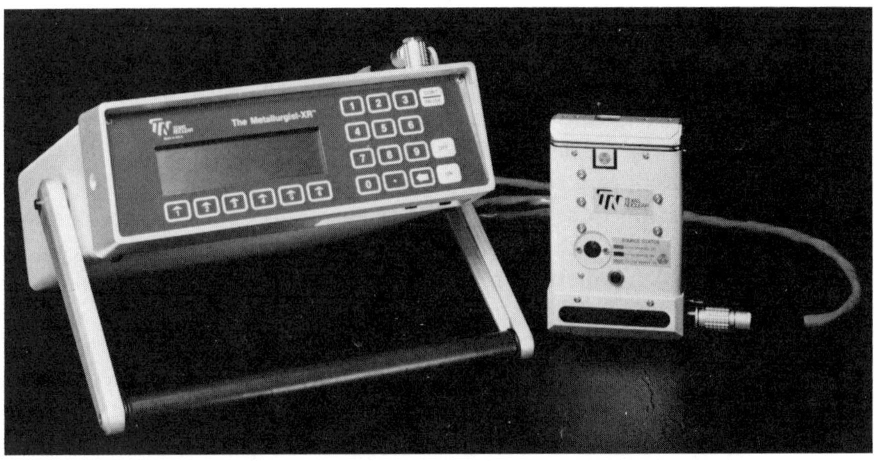

Figure 9-16. XRF metallurgical analyzer: (a) instrument system; (b) typical application. (Courtesy of TN Technologies, Inc.)

9.7 BETA-, GAMMA-, AND X-RAY SPECTROMETERS

In *beta-ray spectrometry,* the energy levels or momentum levels of electrons are analyzed, using one of several types of magnetic-field analyzers or electrostatic analyzers (ESA) in conjunction with an electron detector, usually of the continuous or multiple dynode type. *Gamma-ray* and *X-ray spectrometers* (Figure 9-17) analyze the number of events occurring in a detector (and proportional to intensity of incident electromagnetic radiation) over a range of photon energies (which correlate with wavelength and frequency; see Figure 1-1). Hence, they are spectroradiometers for the spectral region between about 10 and 10^{-4} nm; expressed in photon energy, as is customary in this field of spectrometry, this spectral region extends from 0.1 to 50 keV, for X rays, and then from 50 keV to nearly 10 MeV, for gamma rays.

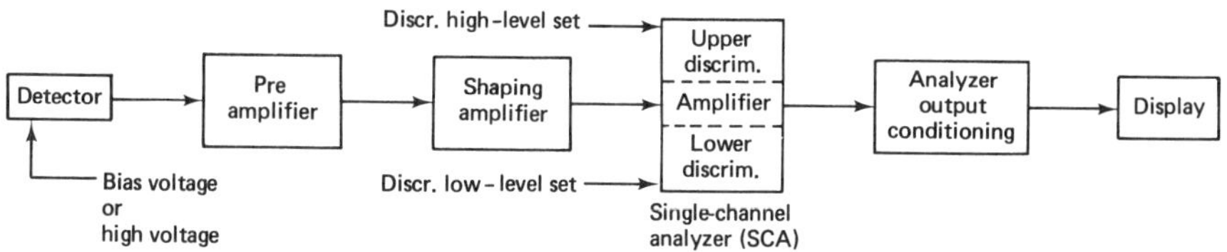

Figure 9-17. Basic block diagram of gamma ray or X-ray spectrometer with SCA.

Detectors used in gamma-ray and X-ray spectrometers are typical nuclear radiation detectors. Proportional counters are useful for ultrasoft to soft X-ray detection, and some semiconductor detectors, notable Si(Li) detectors, which usually have a very thin beryllium window, are suitable for detection of soft to hard X rays. Scintillation counters, typically those using a NaI(Tl) scintillator crystal, are usable over a wide range covering X rays and gamma rays. Semiconductor gamma-ray detectors comprise primarily the cryogenically cooled Ge(Li) and intrinsic (high-purity, HP) germanium detectors.

The device comparable to a monochromator in spectroradiometers for longer wavelengths is the *pulse-height discriminator.* The amplitude of the output pulse from the detector is proportional to the energy of the radiation incident on the detector. By passing the pulse through an amplifier that passes only pulses having amplitudes between two voltage levels, the number of pulses in a narrow range of energies can be counted. The two voltage levels are established by two *discriminators* (gates), one acting as the upper-level discriminator, the other as the lower-level discriminator. The difference between the lower and upper level of energies (that are equivalent to the lower and upper levels of the pulse amplitude) is called the *window;* the window is comparable to the band-pass of a monochromator, and it is expressed in eV or keV.

Two basic types of pulse-height discriminators are used in gamma- and X-ray spectrometers. The *single-channel analyzer* (*SCA*) (see Figure 9-17) contains a single set of upper and lower discriminators. When the voltage settings of the two discriminators (and, hence, the window) are fixed, only pulses whose energy fits into that window will be passed. However, when the voltage settings of the two discriminators are increased continuously (preferably with the ramps being a precise function of time), but with the voltage difference between them (the window) kept the same, the SCA will scan a given spectral region of energies from its lower to its upper limit as the window slides upward. As a result, a spectrum display (Figure 9-19a) can be obtained. Such a spectrum will show peaks whose location on the energy scale enables identification of the radiation-emitting isotope, whereas the net area under each *photopeak* is proportional to the emission rate of the isotope (having its peak at a given energy level). It can also be noted from Figure 9-19a that the peaks do not start at zero but protrude above a background (*Compton continuum*) due to photons losing only part of their energy to the detector. The ordinate (Y-axis) of the graph is usually logarithmic and is graduated in number of pulses (*counts*), per unit time (or obtained over a known time period) per window (i.e., in the window, as it scans increasing photon energies). In calculations of absolute emission rates by area integration, the Compton continuum can be subtracted or corrected for by comparisons to standard curves.

The *multichannel analyzer* (*MCA*) can be compared to an interferometric spectroradiometer in that it analyses all parts of the spectrum essentially at the same time. The MCA (Figure 9-18) contains a large number of individual discriminator sets (*channels*). Each channel has a window in a different (and usually adjacent) location on the abscissa (denoting energy) of the resulting display. The number of counts (per time unit) for each channel (where each

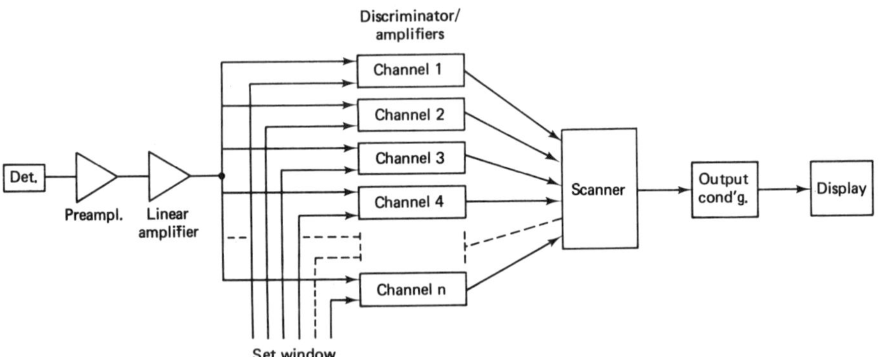

Figure 9-18. Basic block diagram of gamma-ray or X-ray spectrometer with multichannel analyzer (MCA).

channel is equivalent to a very narrow band of energies) is read out sequentially by a high-speed scanner. Modern spectrometers often use digital techniques, particularly when they incorporate an MCA; the appearance of a typical spectrum, as it may appear on a CRT, is illustrated in Figure 9-19b. In MCAs the window is often specified in terms of (the location of) its center and of its width.

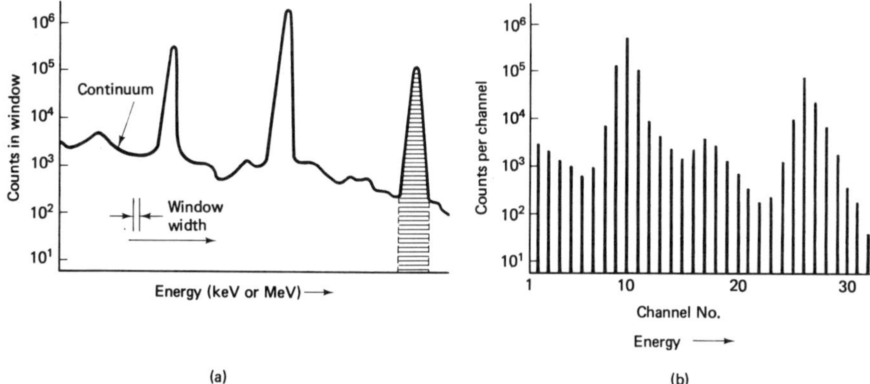

Figure 9-19. Typical spectrum displays (X-rays, keV; gamma rays, MeV): (a) from SCA, strip-chart, analog; (b) from MCA, digital, bar-chart.

Various forms of output conditioning can be made available, again facilitated by use of digital techniques. These may include *scaling* (a *scaler* produces an output pulse for each specified number of input pulses), background removal, smoothing, and differentiation; however, a *logarithmic amplifier* is almost invariably included so that the ordinate of the display can be graduated in powers of ten of counts per channel.

Among the increasing applications of these spectrometers are uses in planetology and astrophysics, for general analytical work on materials, and in *neutron activation analysis*. The latter is used primarily for the analyses of trace impurities in nonradioactive materials. These are irradiated with thermal or fast neutrons and the resulting decay products, beta and gamma rays, are then analyzed by spectrometers.

9.8 SUBMILLIMETER AND MICROWAVE SPECTROMETERS

For applications primarily in the area of remote sensing, recent developments in spectrometry include passive and active (radar) spectrometers used for spectroradiometric analyses of emission and absorption lines. The frequency range of *microwaves* extends from about 1 to 300 GHz (equivalent to wavelengths between 30 cm and 1 mm); the region between about 10 and 1 mm is sometimes called *millimeter waves.* In this region, radio-frequency (*RF*) techniques and instrumentation are generally used. In the region between 1 and 0.1 mm (1000 and 100 μm), which is now often referred to as *submillimeter waves,* there is an overlap between the applicability of far-IR-optical and RF techniques. In these regions, spectrometry is used increasingly to obtain *emission spectra* from primarily stellar and interstellar sources and *absorption spectra* through gases that use as illumination source either radiation of known spectral characteristics, such as solar radiation (in *passive microwave spectrometry*), or pulsed electromagnetic energy emanating from the spectrometer instrument (*active microwave spectrometry, radar spectrometry*).

The incident electromagnetic energy, in microwave spectrometry, is usually gathered by an antenna (*collector*) and fed through a *waveguide* to a *receiver.* Here it is applied to a mixer, where the incoming signal is heterodyned against a local oscillator. The output of the mixer is a band of frequencies of a few hundred MHz, frequencies much easier to work with than the hundreds of GHz of the originally received signal. The separation of the heterodyned signal is accomplished either by a set of parallel filters, each having a narrow and adjacent passband, or by digital techniques requiring use of Fourier transforms for data reduction.

The mixer is probably the most critical element in the receiver. It must be selected and operated so as to minimize conversion loss and any noise introduced by it into the heterodyne (output) signal. Schottky barrier diodes are commonly used as mixers; they are sometimes cooled to reduce noise (the noise is reduced by about 50% when cooled from room temperature to 20 K). Other devices, used as mixers, are still in development and they generally require operation at liquid-helium temperatures; they include Josephson devices, other superconducting devices, and an indium-antimonide bolometer (hot-election bolometer).

10

MAGNETIC RESONANCE SPECTROMETERS

This category of instruments provides spectra from radio-frequency-induced transitions in the presence of a magnetic field. It comprises two types of instruments: nuclear magnetic resonance (NMR) spectrometers and electron spin resonance (ESR) spectrometers. Both are used primarily for research. Additionally, NMR-based scanners are now widely used for medical diagnostics.

10.1 NMR SPECTROMETERS

Many atomic nuclei (about one-half of those known) will exhibit nuclear magnetic resonance when they are exposed to a static field; they will then absorb energy from a radio-frequency field at certain characteristic frequencies. The RF field is applied at right angles to a strong, uniform magnetic field. The resonance occurs when the frequency of the rotating component of the RF field equals the precession frequency of a nucleus that possesses spin or angular momentum in addition to charge and mass. A spinning charge creates a magnetic field. The precession is caused by changes in spin-axis alignment in the presence of the magnetic field. At the point where the nucleus absorbs RF energy it will undergo a transition to a higher energy level.

There are two modes in which an NMR spectrometer can operate. In the *frequency-sweep method,* the magnetic field is kept constant and the output of an oscillator/power-amplifier, coupled into the sample by a coil, is swept over a range of frequencies. A pickup coil is also placed around the sample (orthogonal to the transmitter coil); it is connected to an RF amplifier. When a change in the received and phase-detected signal amplitude indicates an absorption peak, the RF transmitter frequency is recorded. In the *field-sweep method,* the RF transmitter frequency is held constant and the magnetic field is modulated by means of sweep coils around the pole pieces. Resonance is usually detected from a dispersion peak. Figure 10-1 shows a basic block diagram of an NMR spectrometer that can operate in the field-sweep mode.

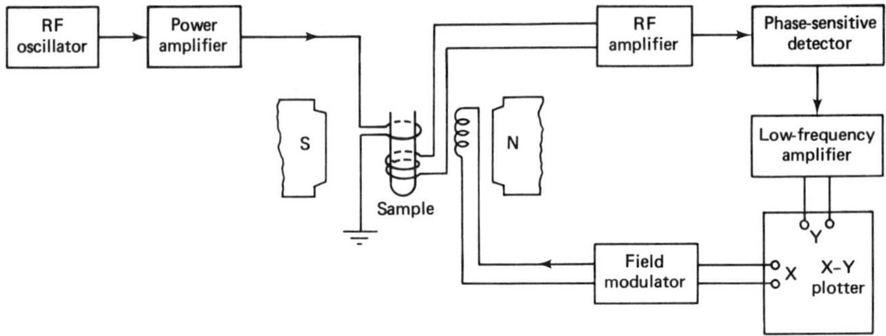

Figure 10-1. Basic block diagram of NMR spectrometer.

The RF equipment must be capable of detecting the resonance-indicating signal within very close frequency tolerances; it must also be capable of very precise frequency control, and provide very accurate knowledge about the frequency of the RF field. The magnetic field must also be held and known within very close limits. The magnetic field is typically in the order of 1 to 3 tesla (T) (10 to 30 kG); however, fields up to 15 T have been obtained using a superconducting magnet. The magnet can be a permanent magnet, or it can be an electromagnet; in the latter case, considerable heat is generated by the magnet coils and the electromagnet is usually water-cooled.

10.2 ESR SPECTROMETERS

Spectrometers whose operation is based on *electron spin resonance,* also known as *electron paramagnetic resonance* (*EPR*), are similar to NMR spectrometers; the main difference is that they operate in the microwave region.

Figure 10-2 shows the approximate regions in which observations are based primarily on nuclear spin magnetism (VHF to UHF), on electron spin magnetism (S-band and, primarily, X-band microwaves), and on molecular rotations (K-band microwaves and millimeter waves).

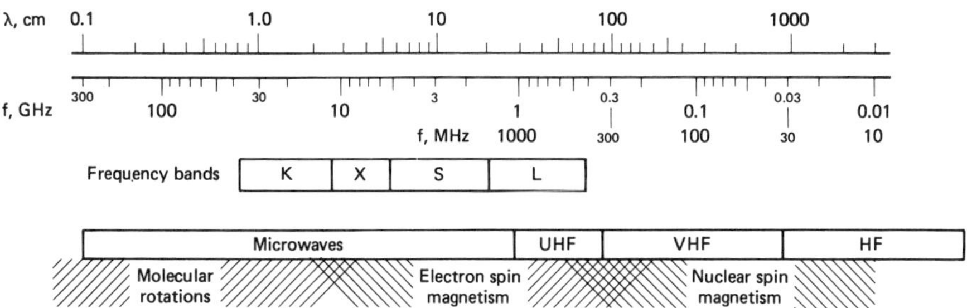

Figure 10-2. Frequency spectrum from high frequencies (HF) through microwaves.

In ESR, radiation (typically around 10 GHz) induces transitions between magnetic energy levels of electrons with unpaired spins, in the presence of a magnetic field. A block diagram of a typical ESR spectrometer is shown in Figure 10-3. The sample is inserted into the resonant cavity, which is located between the poles of an electromagnet. The homogeneous magnetic field produced by this electromagnet can be varied from zero to 0.5 T (0 to

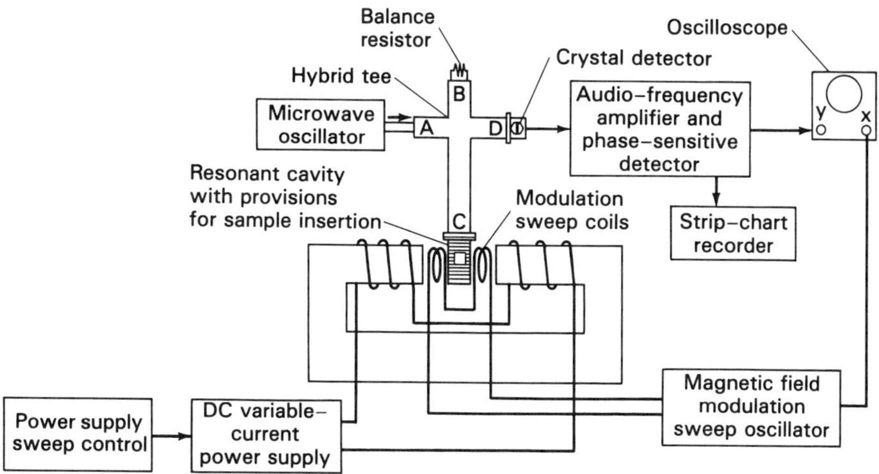

Figure 10-3. Typical ESR spectrometer block diagram.

5000 gauss). In a typical operation the field is swept slowly over a narrow portion of this range where the resonant condition is expected and is additionally fine tuned by modulating the field slightly, using sweep coils powered by an oscillator whose frequency can be varied from the low audio range to the very low RF range.

The instrument illustrated uses a microwave bridge in the form of a *hybrid tee.* A resistive load is used as the termination of arm B to establish bridge balance, so that energy applied to arm A divides equally between arms B and C, and no energy is seen by arm D which contains a crystal detector. When a resonance condition occurs in the cavity (part of arm C), the sample absorbs energy, causing a bridge unbalance, and a signal is then seen by the detector.

The information displayed is either the absorption spectrum or its derivative, or both. The total area enclosed by either is indicative of the number of unpaired electrons in the sample. Typical displays are an oscilloscope (synchronized to the low-frequency sweep oscillator) and a strip-chart recorder for hard copy. ESR spectroscopy is particularly useful in research on odd molecules, free radicals, biradicals, and transition element ions.

11

MASS SPECTROMETERS

11.1 GENERAL OPERATING PRINCIPLES

Mass spectrometers can be used to determine the composition as well as the abundance of constituents of very small samples. The constituents can be elements (atoms), including isotopes, and molecules, including isomers. All mass spectrometers analyze the mass/charge ratio of ions; the ions can be the parent ion or ionic fragments of a molecule. Virtually all these instruments are neutral mass spectrometers and provide an ionizing chamber (ion source) to produce charged ions. A few types, used in research, respond only to incoming charged ions (ion mass spectrometers) and do not contain an ion source.

The peaks appearing in the display (the mass spectrum) show the number of ions having the same mass-to-charge (m/e) ratio. Usually only positive ions are analyzed. From the mass spectrum, which shows abundance vs. mass unit (in amu), many types of information can be extracted, such as the organic-compound structure and constituents of complex mixtures, and the molecular weight can be determined with a high degree of accuracy.

One of the earliest practical mass spectrometers was the *RF mass spectrometer,* in which ions are sorted by passing them through sets of grids;

alternate grids are at a steady potential, whereas the other set of alternate grids is supplied from a radio-frequency (RF) source. The RF frequency is variable; at a given frequency, only ions of a given mass/charge ratio will pass through the grid structure. Most modern mass spectrometers are either of the magnetic-sector or the quadrupole type, but some are of the time-of-flight type; these will be explained below in more detail.

The basic elements of a mass spectrometer are the ion source, the mass analyzer, and the electron detector, with their associated power supply, control, and signal conditioning circuitry, and the display, as illustrated in Figure 11-1. The sample, typically at room pressure and in gaseous form, is introduced into the vacuum analyzer system usually through a porous disc or membrane (molecular leak), after any required conditioning. Mass spectrometers are increasingly used on-line, i.e., with the sample taken directly from a process stream. Solid samples are first vaporized by passing them through a pyrolizer. To ensure the vacuum in the analyzer system, an ion pump is generally connected to the mass analyzer and, in many designs, a vacuum pump (typically a turbomolecular pump) is also connected to the ion source.

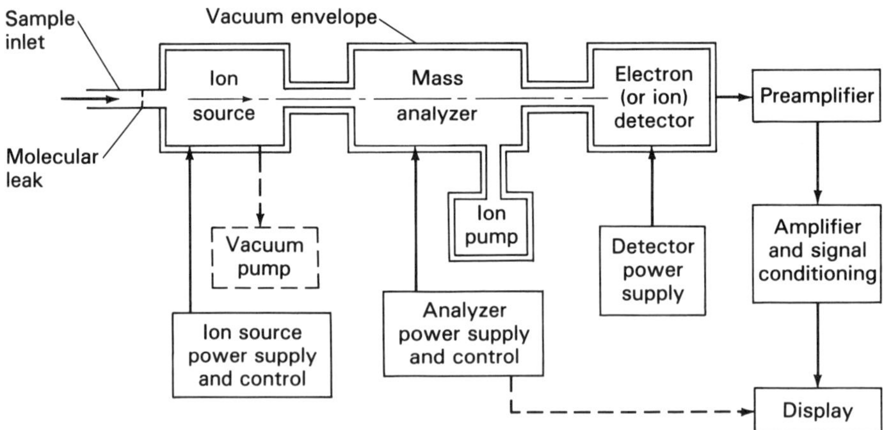

Figure 11-1. Typical mass spectrometer block diagram.

The *ion source* converts the incoming particles into charged ions. The ion source typically uses electron impact for this conversion. A heated filament (most instruments contain two of these, for redundancy) emits electrons that are accelerated toward the anode. The impact (or passage in close proximity) of the electrons ionizes the atoms and molecules of the sample material. A lens-electrode system focuses the ions into a beam sufficiently narrow to enter the mass analyzer through its entrance aperture. Some other types of ion sources have also been employed. They may use a high-density electrostatic field (field ionization, *FI,* or field desorption, *FD*), a high-intensity RF spark, chemical ionization (*CI*), vaporization from a surface heated to high temperatures (about 2000 °C), or a laser beam.

Mass analyzers exist in a variety of generic types and designs; three commonly used types are described below. When they respond to a single beam, electron detectors are usually electron multipliers, sometimes of the multiple dynode type, but more commonly of the continuous dynode type (channel electron multipliers). The operation of the electron detector relies upon the production of secondary electrons when the incident ions strike the detector. Faraday collectors are used in some designs for the direct detection of high-level ion currents. Computer control, including for the display, is increasingly used in modern mass spectrometers.

While mass spectrometers continue to play an important role as laboratory instruments, ruggedized, reliable designs are now increasingly used online, for example, in biotechnology, for characterization of stack emissions, and in many types of chemical process control. For some types of analyses it has been found advantageous to couple a mass spectrometer to a liquid chromatograph (*LC/MS*) or a gas chromatograph (*GC/MS*). Chromatographs are covered in Chapter 8.

11.2 QUADRUPOLE MASS ANALYZERS

A typical quadrupole mass filter (see Figure 11-2) consists of four precision-machined cylindrical rods located in an orthogonal array. Opposite pairs of rods are connected electrically, with a dc voltage (U) as well as an RF voltage (V) applied between the rod pair. Connected in this manner, one pair of rods is negative, the other positive, and the RF voltage across the negative pair is phase-shifted by 180° with respect to the positive pair. As ions move through the field in the gap between the rods, oscillations are induced in their motion. The amplitude of these oscillations depends on the dc voltage as well as the frequency of the RF component. With a given combination of dc voltage and RF frequency, only ions of a given mass-to-charge ratio will stably traverse the mass filter. Other ions (nonresonant ions) will assume trajectories that causes them to strike a rod and be neutralized. Rapid changes of the applied potentials allow quick scanning of the mass range (typically up to 1000 amu). To improve the mass separation of ions with a wide energy spectrum, an electrostatic energy filter can be placed before the mass filter. The energy filter then selects a narrow energy band pass (less than 10 eV) for the secondary ions. Ion sources for quadrupole mass spectrometers can be relatively simple. This type of instrument has moderate mass resolution capability.

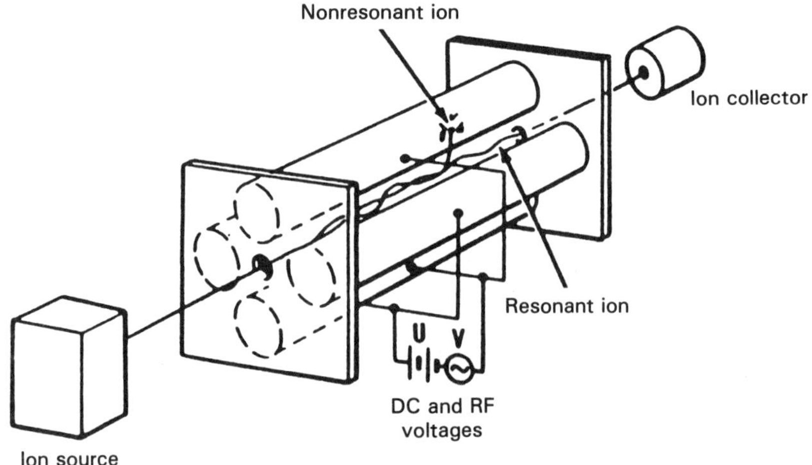

Figure 11-2. Quadrupole mass analyzer. (Courtesy of Perkin-Elmer Corp.—Applied Science Operation.)

11.3 MAGNETIC-SECTOR MASS SPECTROMETERS

Magnetic-sector analyzers are generally of two types: single-focusing or double-focusing. Figure 11-3 illustrates a *single-focusing analyzer.* The beam from the ion source passes through the magnetic sector which applies a transverse magnetic field to the moving ions. The geometry and curvature of the analyzer magnet forces the ions to travel along a curved trajectory, permitting ions of different mass-to-charge ratios to be sorted out on the basis that each different ion will have a different trajectory through the magnetic field. The curvature of the magnetic sector shown is 90°; in other such analyzers the curvature can be 60° (*Nier* system), 180° (*Dempster* system), or 120°.

The forces exerted on the ions are mass-, charge-, and velocity-dependent. The ions are detected by Faraday collectors placed at a set of different radii, and electrometers are employed to produce appropriate electrical output signals. This type of instrument has a mass range of about 2 to 200 amu.

A *double-focusing electrostatic/magnetic sector analyzer* (*DF/MS*) is shown in Figure 11-4. In this instrument, the ions first pass through a curved electrostatic analyzer section, which permits their selection according to energy, then through a straight section, and finally through a curved magnetic sector. For a given ion energy (selected by the electrostatic sector), the radius at which a certain mass is focused to pass through the exit aperture is varied by

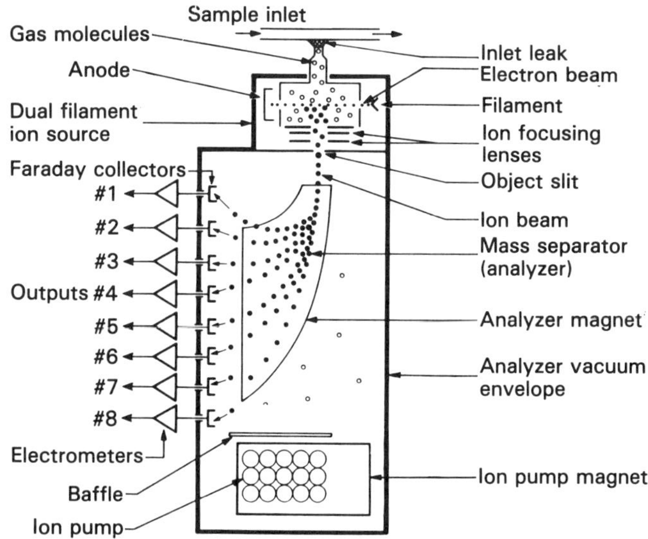

Figure 11-3. Single-focusing magnetic-sector analyser. (Courtesy of Perkin-Elmer Corp—Applied Science Operation.)

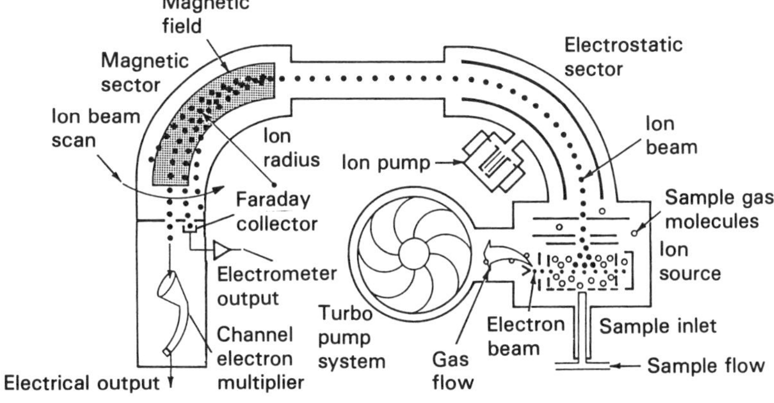

Figure 11-4. Double-focusing analyzer with electrostatic and magnetic sectors. (Courtesy of Perkin-Elmer Corp.—Applied Science Operation.)

changing the magnetic field strength. Since the magnet has some amount of inherent hysteresis, the scanning process is slower than in some other types of mass spectrometers; however, the DF/MS provides very good mass resolution. The mass spectral range usually extends to about 250 amu.

11.4 TIME-OF-FLIGHT MASS SPECTROMETERS

The operation of a time-of-flight (*TOF*) mass spectrometer is based on the difference in velocities acquired by ions of different mass, if they are given the same kinetic energy. The design requires that the ion source be pulsed (typically pulses of 10 ns or less). The pulses of secondary ions, generated when the ion beam strikes the sample, are accelerated to attain a velocity that is a function of their mass/charge ratio. The ions then travel through an evacuated drift tube and strike a large-area detector. The time it takes an ion to reach the detector is determined by its mass. By using an appropriate display device, such as an oscilloscope whose time base is synchronized with and triggered by the pulse generation, the total mass spectrum can be scanned in a few microseconds.

The resolution of a TOF analyzer is comparable to that of the magnetic-sector type. This type of instrument is generally used to detect high molecular weight masses or to perform very rapid analyses. The very brief analysis time is particularly useful in organic analyses, because sample damage is minimized.

12

SURFACE ANALYSIS INSTRUMENTS

12.1 INSTRUMENT BASICS

A large number of different instruments and instrumental methods have become available for the microanalyses of surfaces of many types of materials. Some of those are used primarily to analyze the properties and composition of the topmost surface layer, which may be monomolecular (one *monolayer* thick); others are used to penetrate the surface down to 50 or more monolayers; one of the purposes of such analyses is subsurface *profiling*. The dimensions involved in surface analysis are microscopically small and are typically expressed in nm (or angstroms, 10 Å = 1 nm).

The basic operating principle of surface analysis instruments is the following (see Figure 12-1): the sample whose surface is to be analyzed is bombarded by a beam from a source of electrons, ions or (primarily UV or X-ray) photons; the interaction of this beam produces electrons or ions (or sometimes, photons or neutrons) which are focused into the entrance aperture of an analyzer. For most instruments this is an energy analyzer; in some instruments it is a mass analyzer, or an energy analyzer followed by a mass analyzer. A detector is mounted to the output side of the analyzer. The de-

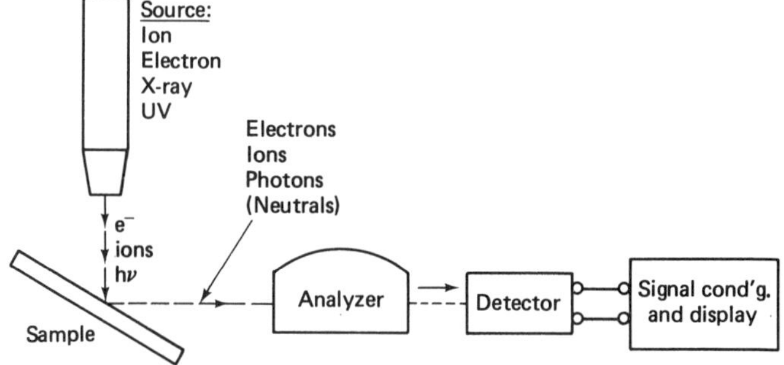

Figure 12-1. Basic elements of surface analysis instruments.

tector is typically an electron multiplier of the multiple-dynode or continuous-dynode (e.g., channel electron multiplier, CEM, or microchannel plate, MCP) type.

Of the many instruments and instrumental methods, only a few of the more frequently used ones will be described very briefly. However, the field of surface analysis has been the birthplace of a large number of acronyms, and an understanding of these is essential to technical communications. The meanings of most of the acronyms used in surface analysis are listed below, in alphabetical order; in most of these, where the acronym pertains to spectroscopy as a method, the word spectrometer can be substituted to denote the instrument used; however, it should be noted that many instruments are usable for more than one instrumental method; hence, it is better to associate the acronym with a method rather than a specific instrument design.

AES auger electron spectroscopy
APS appearance potential spectroscopy
CELS characteristic electron loss spectroscopy
CIS characteristic isochromat spectroscopy
EID electron-induced desorption
EIID electron-impact ion desorption
ELS electron loss spectroscopy
EMP electron microprobe
ESCA electron spectroscopy for chemical analysis (same as XPS)
ESD electron-stimulated desorption (see EID and EIID)
FEM field-emission microscope
HEED high-energy electron diffraction

ILEED	inelastic low-energy electron diffraction
INS	ion neutralization spectroscopy
ISS	ion scattering spectroscopy
LEED	low-energy electron diffraction
LEIS	low-energy ion-scattering spectroscopy
PIX	proton-induced X-ray analysis
PLEED	polarized low-energy electron diffraction
PSD	photon-stimulated desorption
RHEED	reflected high-energy electron diffraction
SALI	surface analysis by laser ionization
SAM	scanning auger microscopy
SEM	scanning electron microscopy
SIMS	secondary ion mass spectroscopy
SLEEP	scanning low-energy electron probe
UPS	ultraviolet photoelectron spectroscopy
XPS	X-ray photoelectron spectroscopy

Sources used for surface analysis include primarily the electron gun, the ion gun, the X-ray source and the ultraviolet (UV) lamp. An *electron gun* (electron source) consists typically of a heated cathode, from the surface of which electrons are emitted, a control grid, an accelerating grid which increases the velocity of the electrons, focusing grids which shape the electrons into a narrow beam, and deflecting electrodes which control the direction of the exiting beam. For most methods, the direction of the beam is held constant. For scanning methods (e.g., SAM, SEM), the potentials applied to the two pairs of electrodes (X- and Y-axis) can be programmed so that the beam scans over the sample surface, usually forming a raster. In an *ion gun* (ion source) an electron source is used to bombard atoms of a gas (e.g., He, Ne, Ar) that is introduced, at vacuum pressures (10^{-3} to 10^{-5} torr), into the ion gun from an external gas supply. For the usual negative-ion source, the ionization chamber is kept at a negative accelerating potential, and an extractive electrode, at a less negative potential, draws ions from the source through a small aperture. Lens electrodes are then used to form the required narrow beam of ions that is directed at the sample. Positive-ion beams can be obtained from alkali metal or sputter sources.

An *X-ray* source is, generally, a chamber in which the thermionic-emission electrons are accelerated toward a target by a high potential applied between cathode and anode. X rays are produced when the electrons collide with the metallic target. Anode shaping (e.g., use of an annular anode) or diffraction from a curved crystal can be used to maximize the flow of X-ray photons to the sample. *UV lamps* typically produce their photons by means of a gas discharge in a lamp having a quartz envelope. Photons from either source are sometimes monochromatized with diffraction gratings (see Section 9.2.2) or crystals.

Analyzer sections of surface analysis instruments are electron or ion energy analyzers in most instruments, and mass analyzers in some instruments. Magnetic energy analyzers are no longer in frequent use and the energy analyzer found in most modern instruments is the *electrostatic analyzer* (*ESA*). Whereas some energy analyzers are nondispersive, most are dispersive and act as energy monochromators. One commonly used type of ESA is the *hemispherical analyzer* (Figure 12-2a). The electron beam from the sample enters the analyzer through a lens-electrode system (which may also act so as to apply a retarding field to the electron stream). The analyzer consists of two hemispherically curved parallel plates across which a potential is applied. The higher-energy electrons will follow a trajectory through the analyzer with a larger radius than the lower-energy electrons. By selecting the appropriate electrode potential, the trajectory of electrons having a specified kinetic energy will be exactly that trajectory that causes the electron stream to pass through the exit slit and onto the detector. The detector, usually an electron multiplier, produces an output current proportional to the number of electrons detected per unit time; good electron multipliers, with appropriately designed circuitry, can produce a measurable output pulse for each electron detected.

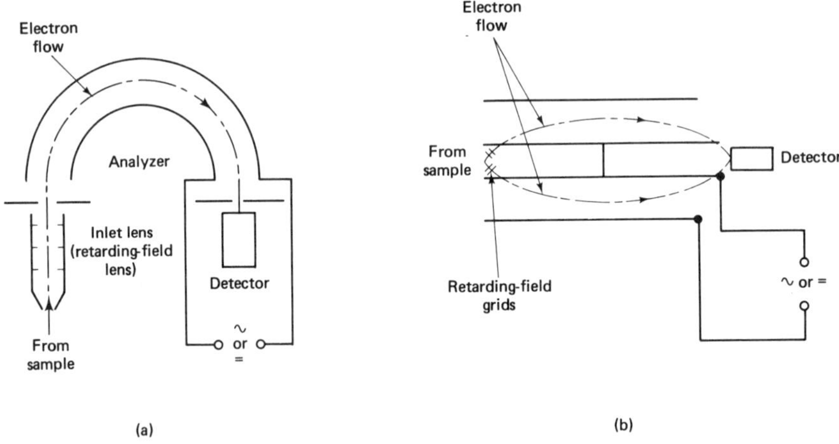

Figure 12-2. Energy analyzers: (a) hemispherical analyzer; (b) cylindrical mirror analyzer (CMA).

The energy of an electron passing through the analyzer and through the exit slit can be determined from knowledge of the analyzer geometry and the applied potential. For a hemispherical analyzer the kinetic energy of the electron, *KE* (in electron volts), is proportional to the potential applied across the electrodes (V) and the mean radius of the two plates (r), and inversely proportional to twice the separation (d) between the plates, or $KE = Vr/2d$. When the energy of ions, rather than of electrons, is to be determined, the result must be divided by the charge of the ion, q, and the relationship is then $KE/q = Vr/2d$.

Another popular ESA design is the *cylindrical mirror analyzer* (*CMA*). As shown in Figure 12-2b, electrons enter the analyzer through a set of retarding-field grids. The analyzer consists of an inner cylinder, which has two annular apertures (electrically conductive) and an outer cylinder. The potential is applied across the two cylinders. The annular apertures form the entrance and exit slits. With a given potential applied, only electrons having the corresponding energy to pass through the slits will have a trajectory that reaches the detector. The usual manner of producing an energy spectrum is to apply a scanning potential across the analyzer electrodes. Another method involves applying a scanning potential to the retarding field grids and keeping the analyzer electrode potential difference at a fixed value.

Mass analyzers are of the types used in mass spectrometers (see Chapter 11). Analyzers used in such applications as SIMS are now usually of the quadrupole type; the ions pass through the line of symmetry between four parallel cylindrical rods; an alternating potential superimposed on a steady potential between pairs of rods, 180° apart, filters out all ions except those of a specific mass, as determined by the applied potentials. This type of analyzer is also known as *mass filter.*

Surface analysis instruments can be classified by the type of source employed. Depending on the energy of the beam from the source, its angle of incidence with the sample surface, the nature of the surface itself, and the source-sample-analyzer geometry, various interactions occur at the sample surface or in subsurface layers that cause either the same or different types of particles to be emitted. Analysis of the particles from the sample can provide information about the nature, structure, composition, and so on, of the sample. It should be noted that all elements of these instruments are maintained in a vacuum, typically around 10^{-10} torr, with much higher vacuums required in some cases. Use of an *electron source* permits analyses based primarily on electrons being emitted from the surface, and the analyses can involve counting the number of electrons, their energy, their spatial distribution, or two or all three of these parameters. In LEED, electrons which are diffracted by the surface from a beam of low-energy electrons (20 to 200 eV) are analyzed; a higher-energy beam is employed in the less frequently applied HEED method. Reflection from the surface, when the source beam is at glancing incidence, is used in RHEED. In ILEED, only electrons that have lost energy by inelastic collisions are analyzed.

One of the most widely used methods employing an electron beam is auger electron spectroscopy (AES). The high-energy electron beam ionizes substrate atoms, which then relax by a radiationless transition. An outer electron drops into the ionized level and transfers the absorbed energy to another electron (*auger electron*) which then leaves the sample and is analyzed for its energy. The electron beam used in AES must have an energy of 2 to 3 keV. A scanning electron beam is employed for rastering in the scanning auger microscope (SAM). Particles emitted from a surface in response to an electron beam may also be desorbed ions (EIID) or neutrals (EID), or photons.

Ion sources are used primarily in ISS and SIMS. In ISS, the energies of ions scattered through a fixed scattering angle are analyzed. In secondary ion mass spectroscopy (SIMS), the ion beam is used to sputter atoms or molecules from the sample surface. The mass of these sputtering products (secondary ions) is then analyzed. In *energy-resolved SIMS* (see block diagram, Figure 12-3), the energy of the secondary ions is additionally analyzed.

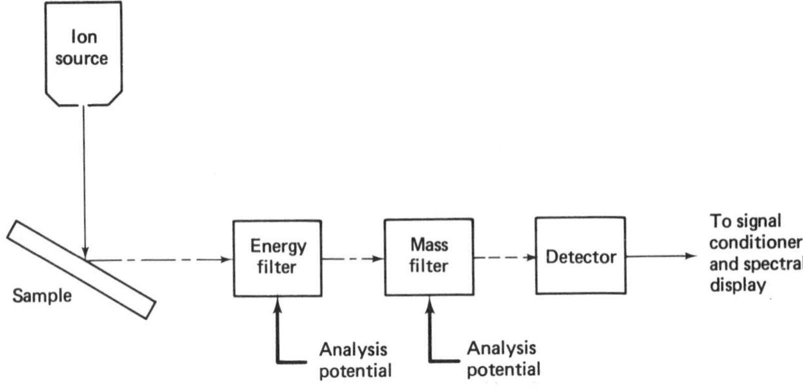

Figure 12-3. Secondary-ion mass spectrometer (SIMS) (energy-resolved) block diagram.

Photon sources considered here are two types: UV and X ray. Both are used in interaction with surfaces that produce electrons whose energy is then analyzed. Depending on the source used, the instrumental method is either UV or X-ray photoelectron spectroscopy (UPS or XPS). The term "electron spectroscopy for chemical analysis" (ESCA) is used synonymously with XPS.

12.2 IMPLEMENTATION EXAMPLES

Commercial equipment is available for surface and thin film analysis. Such equipment is also undergoing improvements to increase capabilities as industries applications expand, notably in semiconductor research and development. The examples of commercially available hardware described here focus on the most popular types of equipment. A SAM system, which can also operate in an SEM mode, perform AES, and has provisions for optional components that allow SIMS to be performed. Additionally, components of an XPS/ESCA system are described.

Figure 12-4 shows a complete SAM equipment setup. At the left is the complete sample introduction and analysis system assembled to a vacuum console. Located behind this console is the vacuum rough pump. The console at the right contains the electronics rack, control electronics, and the system computer with its display devices.

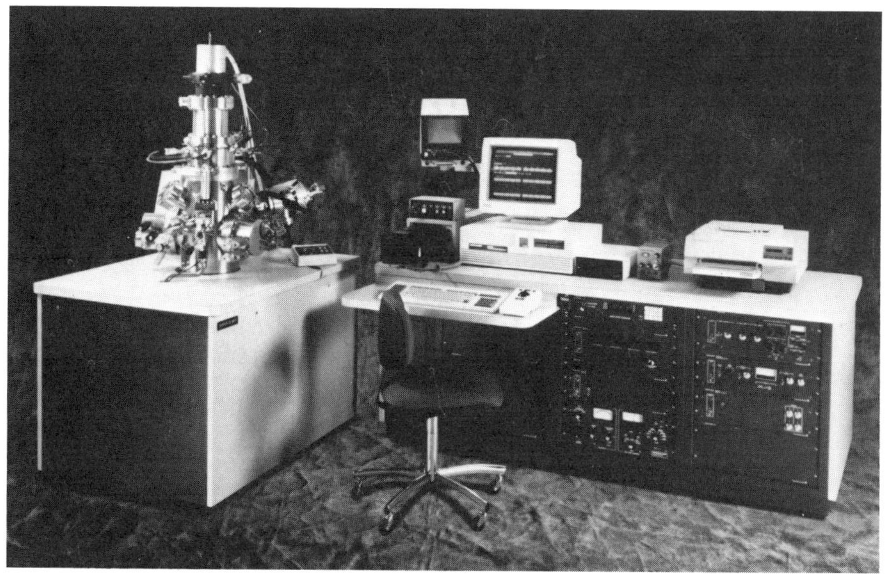

Figure 12-4. Scanning Auger Microscopy (SAM) equipment set-up. (Courtesy of Perkin-Elmer, Physical Electronics Division.)

The major functions and elements of the SAM system are illustrated in the block diagram in Figure 12-5. The diagram includes all the electronic control functions (CMA is the cylindrical mirror analyzer, part of the electron optics). Surface analysis must always be performed in an ultra-high vacuum

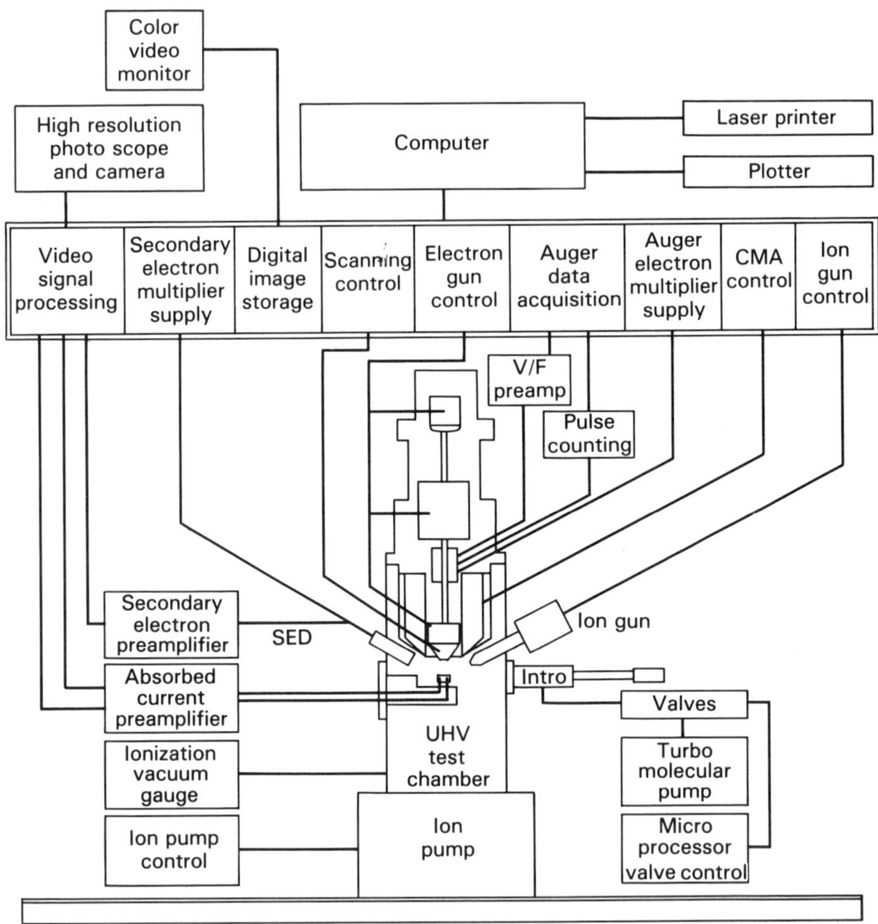

Figure 12-5. SAM system block diagram. (Courtesy of Perkin-Elmer, Physical Electronics Division.)

(UHV), on the order of 10^{-10} torr; UHV conditions significantly extend the time that a sample can be analyzed before it is covered with atoms. The rough pump can only provide a moderate vacuum. The primary UHV pumping is accomplished by high-capacity (220 L/s) ion pumps (seen at the bottom of the diagram). An ionization vacuum gage is used to monitor the vacuum. Heating elements are built into the vacuum console to allow bake-out (high-temperature decontamination of internal surfaces). Vacuum interlocks protect optics components against vacuum failure. A turbo-molecular pump is connected to the specimen introduction assembly to pump out the introduction chamber, which is automatically backfilled with nitrogen during specimen insertion and removal. A mechanical stop is used to assure the specimen stage is placed at the desired location. The specimen stage, itself, allows the specimen to be moved along the *X*-, *Y*-, and *Z*-axes. Optional modules for this stage further allow rotation, heating, and cooling of the specimen.

The ion gun is used to produce a beam for inert ion bombardment of the specimen that results in its sputter etching. This technique is used for thin-film analysis. For the (optional) SIMS operation of the system, a different ion gun is used and a SIMS detector added. The secondary electron detector (SED), with its closely-coupled preamplifier, is used for operation of the system in the SEM (Scanning Electron Microscopy) mode. Electron beam scanning control is digital; the system can produce auger electron spectra, can examine a tiny area of the specimen, scan across a line, or scan over a large number of lines to produce a raster used for imaging. Display devices are included for all of these options and modes. In auger electron spectra, the auger peaks are evident in the total energy distribution function N(E); peak energy is identified by, and elemental composition can be inferred from, the maximum *negative* excursions in the dN(E)/dE peaks (plotted versus kinetic energy, expressed in eV).

The "skeleton" of all surface analysis systems is a vacuum chamber provided with flanged ports, arranged in a specific geometry, and with ports and flange diameters of sizes appropriate to the component or subsystem to be installed in them. All flanges provide for an UHV seal when the bolts around their bolt circle are tightened. Stainless steel is typically used for the chamber, ports, and flanges, as well as for mating flanges and housings of components and subsystems. The geometry and port assignments of the vacuum chamber used in the SAM system is shown in Figure 12-6. Different

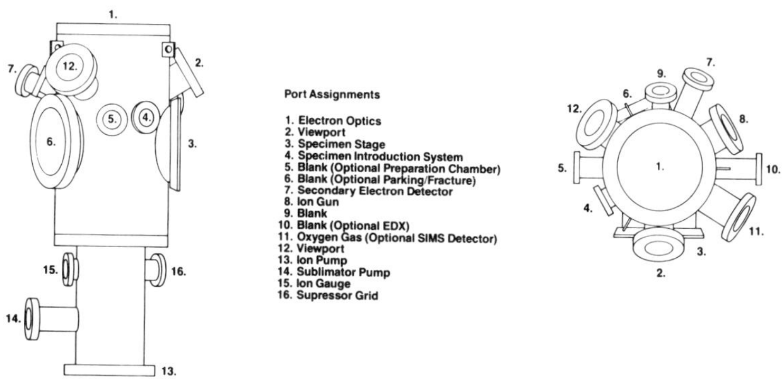

Figure 12-6. Vacuum chamber with flanged ports for SAM system. (Courtesy of Perkin-Elmer, Physical Electronics Division.)

geometries and port locations (and angles) are used in other types of surface analysis equipment; however, the basically cylindrical configuration is typical.

The electron optics, whose major elements are the electron gun and the electron energy analyzer (cylindrical mirror analyzer, CMA, Figure 12-2b), are illustrated in Figure 12-7. The geometry of these two elements is coaxial, with the relatively large CMA aperture surrounding the electron beam exit. This coaxial arrangement nearly eliminates shadowing effects and facilitates accurate sample alignment.

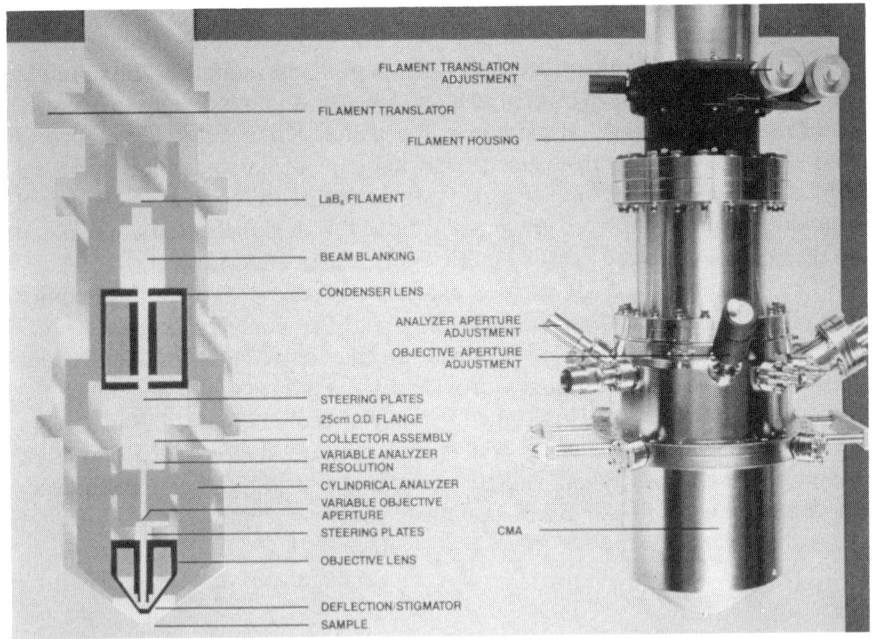

Figure 12-7. Electron optics for SAM system. (Courtesy of Perkin-Elmer, Physical Electronics Division.)

Figure 12-8 shows the "skeleton", i.e., the vacuum chamber with flanged ports, of an ESCA/XPS system. This arrangement is quite different from that of the SAM system in Figure 12-6. The essential elements of this system are an X-ray source, which focuses an X-ray beam onto the specimen, and an electron analyzer that receives the resulting electrons from the specimen, plus, of course, the specimen handling subsystem and the control and display equipment.

This system employs a spherical capacitor analyzer (hemispherical analyzer, Figure 12-2a) as its electron optics (see Figure 12-9). An X-ray source,

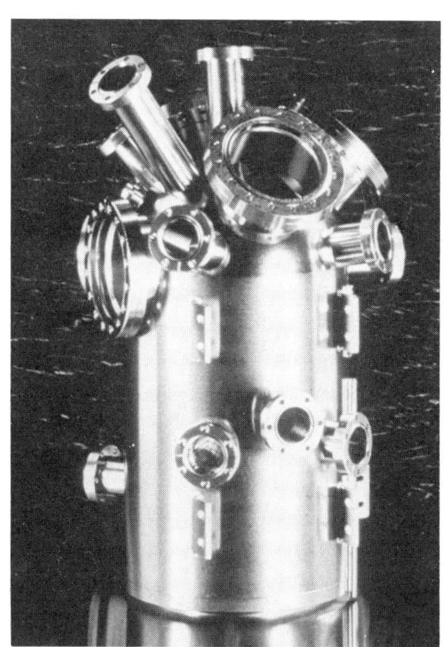

Figure 12-8. Vacuum chamber with flanged ports for ESCA system. (Courtesy of Perkin-Elmer, Physical Electronics Division.)

Figure 12-9. Hemispherical analyzer (spherical capacitor analyzer, SCA) for ESCA system. (Courtesy of Perkin-Elmer, Physical Electronics Division.)

similar to the one used in this system, is illustrated in Figure 12-10. The X-ray source is associated with a crystal-type monochromator designed to provide a very narrow X-ray energy distribution band. The source uses a dual anode design for increased X-ray intensity at the specimen. The anodes are easily interchangeable. Mg and Al anodes are most widely used for ESCA.

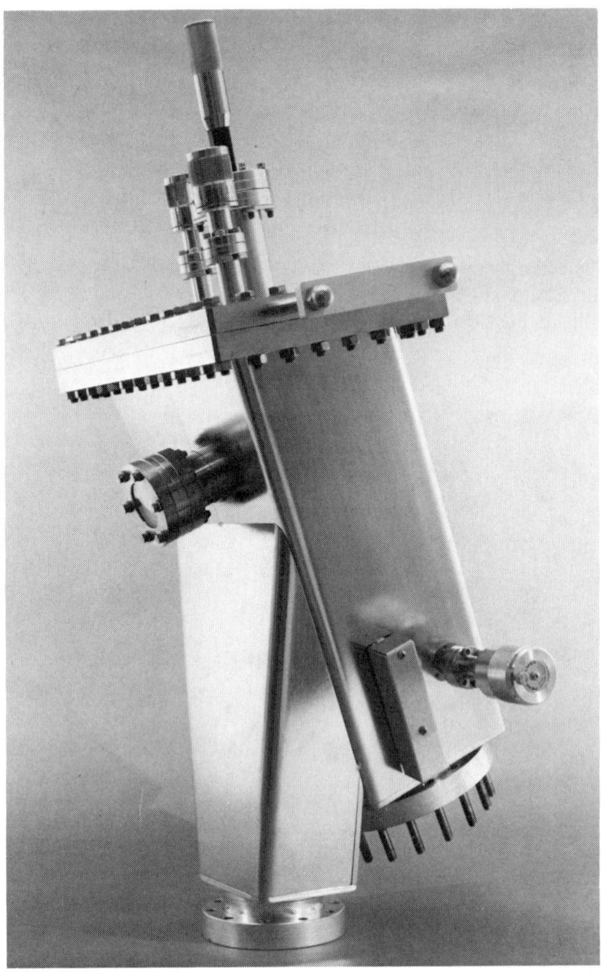

Figure 12-10. X-ray source for XPS. (Courtesy of Perkin-Elmer, Physical Electronics Division.)

BIBLIOGRAPHY

GENERAL

VERDIN, A. *Gas Analysis Instrumentation.* New York: John Wiley & Sons, 1973.

DICK, J.G. *Analytical Chemistry.* New York: McGraw-Hill, 1973.

WILLARD, H.H., MERRITT, L.L., JR., AND DEAN, J.A. *Instrumental Methods of Analysis.* New York: Van Nostrand, 1974.

LINTZ, J., AND SIMONETT, D. *Remote Sensing of Environment.* Reading, MA: Addison-Wesley, 1976.

MOWERY, R.A., JR. *Automated Stream Analysis for Process Control.* New York: Academic Press, 1982.

COLWELL, R.N., ED. *Manual of Remote Sensing.* Falls Church, VA: American Society of Photogrammetry, 1983.

MAGISON, E.C., AND CALDER, W. *Electrical Safety in Hazardous Locations.* Research Triangle Park, NC: Instrument Society of America, 1983.

CLEVETT, K.J. *Process Analyzer Technology.* New York: John Wiley & Sons, 1986.

ELACHI, C. *Introduction to the Physics and Techniques of Remote Sensing.* New York: John Wiley & Sons, 1987.

Deshpande, P.B., and Ash, R.H. *Computer Process Control with Advanced Control Applications*. Research Triangle Park, NC: Instrument Society of America, 1988.

Strock, O.J. *Telemetry Computer Systems: The New Generation*. Research Triangle Park, NC: Instrument Society of America, 1988.

Handbook of Chemistry and Physics. Cleveland: CRC Press, 1991.

ELECTROMETRIC ANALYSIS INSTRUMENTS

Meites, L. *Polarographic Techniques*. 2nd ed. New York: Wiley-Interscience, 1966.

Milner, G.W.C., and Phillips, G. *Coulometry in Analytical Chemistry*. London: Pergamon Press, 1968.

Browning, D.R., ed. *Electrometric Methods*. Maidenhead, England: McGraw-Hill, 1969.

Durst, R.A., ed. "Ion Selective Electrodes." *NIST (NBS) Special Publication 314*. Washington: U.S. Government Printing Office, 1969.

Smith, D.E., and Zimmerli, F.H. *Electrochemical Methods of Process Analysis*. Research Triangle Park, NC: Instrument Society of America, 1972.

Verdin, A. *Gas Analysis Instrumentation*. London: Macmillan, 1973.

Hitchman, M.L. *Measurement of Dissolved Oxygen*. New York: John Wiley & Sons, 1978.

McMillan, G.K. *pH Control*. Research Triangle Park, NC: Instrument Society of America, 1988.

THERMAL ANALYSIS INSTRUMENTS

Wendlandt, W.W. *Thermal Methods of Analysis*. New York: Wiley-Interscience, 1964.

Garn, P.D. *Thermoanalytical Methods of Investigation*. New York: Academic Press, 1965.

Mackenzie, R.C., ed. *Differential Thermal Analysis*. London: Academic Press Ltd., 1970.

CHROMATOGRAPHS

Purnell, H. *Gas Chromatographs*. New York: Wiley-Interscience, 1968.

Jones, R.A. *An Introduction to Gas-Liquid Chromatography*. New York: Academic Press, 1970.

Littlewood, A.B. *Gas Chromatography: Principles, Techniques, and Applications*. 2nd ed. New York: Academic Press, 1970.

Snyder, L.R., and Kirkland, J.J. *Introduction to Modern Liquid Chromatography*. 2nd ed. New York: John Wiley & Sons, 1974.

SCOTT, R.P.W. *Contemporary Liquid Chromatography.* New York: Wiley-Interscience, 1976.

PHOTOMETRIC AND SPECTRORADIOMETRIC (UV-VIS-IR) ANALYSIS INSTRUMENTS

CLARK, G.L., ED. *The Encyclopedia of Spectroscopy.* New York: Van Nostrand Reinhold, 1960.

WRIGHT, W.D. *The Measurement of Colour.* London: Hilger & Watts, 1963.

FREEMAN, S.K., ED. *Interpretive Spectroscopy.* New York: Van Nostrand Reinhold, 1965.

HERCULES, D.M. *Fluorescence and Phosphorescence Analysis.* New York: Wiley-Interscience, 1966.

WYSZECKI, G., AND STILES, W.S. *Color Science: Concepts and Methods, Quantitative Data and Formulas.* New York: John Wiley & Sons, 1967.

GUILBAULT, G.G., ED. *Fluorescence.* New York: Marcel Dekker, 1967.

HARRICK, N.J. *Internal Reflection Spectroscopy.* New York: John Wiley & Sons, 1967.

ZANDER, M. *Phosphorimetry.* New York: Academic Press, 1968.

CALDER, A.B. *Photometric Methods of Analysis.* New York: American Elsevier, 1969.

KERKER, M. *The Scattering of Light.* New York: Academic Press, 1969.

POUCHERT, C.J. *The Aldrich Library of Infrared Spectra.* Milwaukee: Aldrich Chemical Co., Inc., 1970.

HIRAYAMA, K. *Handbook of UV and Visible Absorption Spectra of Organic Compounds.* New York: Plenum, 1971.

COOK, B.W., AND JONES, K. *A Programmed Introduction to Infrared Spectroscopy.* London: Heydon & Sons, 1972.

GROVE, E., ED. *Analytical Emission Spectroscopy.* New York: Marcel Dekker, 1971 (Part I), 1972 (Part II).

NIMEROFF, I., ED. "Precision Measurement and Calibration Colorimetry." *NIST (NBS) Special Publication 300, Vol. 2.* Washington: U.S. Government Printing Office, 1972.

WINEFORDNER, J.D., SCHULMAN, S.G., AND O'HAVER, T.C. *Luminescence Spectrometry in Analytical Chemistry.* New York: Wiley-Interscience, 1972.

GRIFFITHS, P.R. *Chemical Fourier Transform Infrared Spectroscopy,* New York: Wiley-Interscience, 1975.

PEARSE, R.W.B., AND GAYDON, A.G. *The Identification of Molecular Spectra.* New York: Halstead Press, 1976.

YU, F.T.S. *Optics and Information Theory.* New York: Wiley-Interscience, 1976.

MATTSON, J.S., MARK, H.B., JR., AND MAC DONALD, H.C., JR., EDS. *Inrared Correlation and Fourier Transform Spectroscopy.* New York: Marcel Dekker, 1977.

WOLFE, W.L., AND ZISSIS, G.J., EDS. *The Infrared Handbook.* Ann Arbor, MI: Infrared Information and Analysis Center, 1978.

VAN DER MAAS, J.H. *Basic Infrared Spectroscopy.* 2nd ed. London: Heyden & Sons, 1979.

NAFIE, L.A., AND VIDRINE, D.W. *Fourier Transform IR Spectroscopy.* New York: Academic Press, 1982.

RAMAN AND LASER SPECTROMETRY

GILSON, T.R., AND HENDRA, P.J. *Laser Raman Spectroscopy.* London: Wiley-Interscience, 1970.

TOBIN, M.C. *Laser Raman Spectroscopy.* New York: Wiley-Interscience, 1971.

COLTHUP, N.B., DALY, L.H., AND WIBERLEY, S.E. *Introduction to Infrared and Raman Spectroscopy.* 2nd ed. New York: Academic Press, 1975.

LONG, D.A. *Raman Spectroscopy.* New York: McGraw-Hill, 1977.

FLAME EMISSION AND ATOMIC ABSORPTION SPECTROSCOPY

DEAN, J.A. *Flame Photometry.* New York: McGraw-Hill, 1960.

MAVRODINEANU, R., AND BOITEUX, H. *Flame Spectrometry.* New York: John Wiley & Sons, 1965.

ELWELL, W.T., AND GIDLEY, J.A.F. *Atomic Absorption Spectrophotometry.* 2nd ed. Elmsford, NY: Pergamon Press, 1966.

REYNOLDS, R.J., ALDOUS, K., AND THOMPSON, K.C. *Atomic Absorption Spectroscopy.* New York: Barnes & Noble, 1970.

DEAN, J.A., AND RAINES, T.C., EDS. *Flame Emission and Atomic Absorption Spectrometry.* New York: Marcel Dekker, 1969 (Vol. 1), 1971 (Vol. 2), 1974 (Vol. 3).

KIRKBRIGHT, G.F., AND SARGENT, M. *Atomic Absorption and Fluorescence Spectroscopy.* New York: Academic Press, 1975.

X-RAY SPECTROMETRY

BIRKS, L.S. *X-Ray Spectrochemical Analysis.* New York: Wiley-Interscience, 1959.

BERTIN, E.P. *Principles and Practice of X-Ray Spectrometric Analysis.* New York: Plenum, 1970.

HERGLOTZ, H.K., AND BIRKS, L.S., EDS. *X-Ray Spectrometry; Practical Spectroscopy Series.* Vol. 2. New York: Marcel Dekker, 1978.

MICROWAVE SPECTROMETRY

INGRAM, D.J.E. *Radio and Microwave Spectroscopy.* New York: Halstead, 1976.

VARMA, R., AND HRUBESH, L.W. *Chemical Analysis by Microwave Spectroscopy.* New York: John Wiley & Sons, 1979.

MAGNETIC RESONANCE SPECTROMETERS

EMSLEY, J.W., FEENEY, J., AND SUTCLIFFE, L.H. *High-Resolution Nuclear Magnetic Resonance Spectroscopy.* Elmsford, NY: Pergamon Press, 1966.

ALGER, R.S. *Electron Paramagnetic Resonance, Techniques and Applications.* New York: Wiley-Interscience, 1968.

JACKMAN, L.M., AND STERNHALL, S. *Applications of NMR Spectroscopy in Organic Chemistry.* 2nd ed. New York: Pergamon Press, 1969.

ABRAHAM, R.J. *Analysis of High-Resolution NMR Spectra.* New York: American Elsevier, 1971.

FARRAR, T.C., AND BECKER, E.D. *Pulse and Fourier Transform NMR.* New York: Academic Press, 1971.

KEVAN, L., AND KISPERT, L.D. *Electron Spin Double Resonance Spectroscopy.* New York: Wiley-Interscience, 1976.

MASS SPECTROMETERS

DAWSON, P.H. *Quadrupole Mass Spectrometry and Its Applications.* New York: Elsevier Scientific Publishing Co., 1976.

GROSS, M.L., ED. *High Performance Mass Spectrometry: Chemical Applications.* Washington: ACS, 1978.

WHITE, F.A., AND WOOD, G.M. *Mass Spectrometry.* New York: John Wiley & Sons, 1986.

SURFACE ANALYSIS INSTRUMENTS

SHIRLEY, D.A. *Electron Spectroscopy.* Amsterdam: North Holland Publishing Co., 1972.

BOTTOMS, W.R., ED. *Scanning Electron Microscopy.* New York: American Elsevier, 1975.

CZANDERNA, A.W., ED. *Methods of Surface Analysis.* New York: American Elsevier, 1975.

VALYI, L., *Atom and Ion Sources.* London: John Wiley and Sons, 1977.

BARR, T.L., AND DAVIS, L.E., EDS. *Applied Surface Analysis (STP 699).* Philadelphia: American Society for Testing and Materials, 1980.

BENNINGHOVEN, A., ED. *Secondary Ion Mass Spectrometry SIMS II-V.* New York: Springer, 1985.

APPENDICES

THE ELEMENTS

TABLE A-1. The Elements: Alphabetical Listing

Notes:

1. Melting and boiling points of some of the less commonly used elements are average values; (S) = sublimation point.
2. Numbers shown in parentheses are estimated values.
3. Density values are shown for the most common form of an element; for solids (and liquids when at 20°C), they are equivalent to the specific gravity usually at 20°C, in a few cases at a slightly higher or lower temperature; the density of gases is given at 1 atmosphere and 0°C; in some cases the density is also shown for the liquefied state, denoted by L. Density is not shown for some of the radioactive elements.

Element	Symbol	At. No.	Density (g/cm^3)	Melting Pt. (°C)	Boiling Pt. (°C)
Actinium	Ac	89	10.07	1050	3200
Aluminum	Al	13	2.7	660.37	2467
Americium	Am	95	13.67	994	2607
Antimony	Sb	51	6.69	630.74	1950
Argon	Ar	18	1.78 E-3	−189.2	−185.7

Element	Symbol	At. No.	Density (g/cm^3)	Melting Pt. (°C)	Boiling Pt. (°C)
Arsenic	As	33	5.73	817(@ 28 atm)	613(S)
Astatine	At	85	—	302	(337)
Barium	Ba	56	3.5	725	1640
Berkelium	Bk	97	(14)	—	—
Beryllium	Be	4	1.85	1278	2970
Bismuth	Bi	83	9.75	271.3	1560
Boron	B	5	2.35	2079	2550(S)
Bromine	Br	35	3.12	−7.2	58.8
Cadmium	Cd	48	8.65	320.9	765
Calcium	Ca	20	1.55	839	1484
Californium	Cf	98	—	—	—
Carbon	C	6	1.9	—	3367(S)
Cerium	Ce	58	6.66	799	3426
Cesium	Cs	55	1.87	28.4	669.3
Chlorine	Cl	17	3.21 E-3 L: 1.56	−100.98	−34.6
Chromium	Cr	24	7.19	1857	2672
Cobalt	Co	27	8.9	1495	2870
Copper	Cu	29	8.96	1083.4	2567
Curium	Cm	96	13.5	1340	—
Dysprosium	Dy	66	8.55	1412	2562
Einsteinium	Es	99	—	—	—
Erbium	Er	68	9.07	159	2863
Europium	Eu	63	5.24	822	1597
Fermium	Fm	100	—	—	—
Fluorine	F	9	1.7 E-3 L: 1.1	−219.62	−188.14
Francium	Fr	87	—	27	677
Gadolinium	Gd	64	7.9	1313	3266
Gallium	Ga	31	5.9	29.78	2403
Germanium	Ge	32	5.32	937.4	2830
Gold	Au	79	19.3	1064.43	3080
Hafnium	Hf	72	13.3	2227	4602
Helium	He	2	1.78 E-4 L: 0.12	−272.2	−268.9
Holmium	Ho	67	8.79	1474	2695
Hydrogen	H	1	8.99 E-5 L: 0.071	−259.14	−252.87
Indium	In	49	7.31	156.61	2080
Iodine	I	53	4.93	113.5	184.35
Iridium	Ir	77	22.42	2410	4130
Iron	Fe	26	7.87	1535	2750
Krypton	Kr	36	3.73 E-3	−156.6	−152.3
Lanthanum	La	57	6.15	921	3457
Lawrencium	Lr	103	—	—	—
Lead	Pb	82	11.35	327.50	1740
Lithium	Li	3	0.53	180.54	1342
Lutetium	Lu	71	9.84	1663	3395
Magnesium	Mg	12	1.74	648.8	1090

Manganese	Mn	25	7.3	1244	1962
Mendelevium	Md	101	—	—	—
Mercury	Hg	80	13.55	−38.84	356.58
Molybdenum	Mo	42	10.22	2617	4612
Neodymium	Nd	60	6.8	1021	3068
Neon	Ne	10	9.0 E-3 L: 1.21	−248.67	−246.05
Neptunium	Np	93	20.25	640	(3902)
Nickel	Ni	28	8.9	1453	2732
Niobium	Nb	41	8.57	2468	4742
Nitrogen	N	7	1.25 E-3 L: 0.81	−209.86	−195.8
Nobelium	No	102	—	—	—
Osmium	Os	76	22.57	3045	5027
Oxygen	O	8	1.43 E-3 L: 1.14	−218.4	−182.96
Palladium	Pd	46	12.02	1554	2970
Phosphorus	P	15	1.82(Wh)	44.1	280
Platinum	Pt	78	21.45	1772	3827
Plutonium	Pu	94	19.84	641	3232
Polonium	Po	84	9.32	254	962
Potassium	K	19	0.86	63.25	760
Praseodymium	Pr	59	6.77	931	3512
Promethium	Pm	61	7.22	1168	2460
Protactinium	Pa	91	(15.4)	1596	—
Radium	Ra	88	(5.0)	700	1140
Radon	Rn	86	9.73 E-3 L: 4.4	−71	−61.8
Rhenium	Re	75	21.02	3180	(5627)
Rhodium	Rh	45	12.41	1966	3727
Rubidium	Rb	37	1.53	38.89	686
Ruthenium	Ru	44	12.41	2310	3900
Samarium	Sm	62	7.52	1077	1791
Scandium	Sc	21	2.99	1541	2831
Selenium	Se	34	4.79	217	685
Silicon	Si	14	2.33	1410	2355
Silver	Ag	47	10.50	961.93	2212
Sodium	Na	11	0.97	97.8	882.9
Strontium	Sr	38	2.54	769	1384
Sulfur	S	16	2.07	112.8	444.67
Tantalum	Ta	73	16.65	2996	5425
Technetium	Tc	43	(11.5)	2172	4877
Tellurium	Te	52	6.24	449.5	990
Terbium	Tb	65	8.23	1356	3123
Thallium	Tl	81	11.85	303.5	1457
Thorium	Th	90	11.72	1756	4790
Thulium	Tm	69	9.32	1545	1947
Tin	Sn	50	7.3	231.97	2270
Titanium	Ti	22	4.54	1660	3287

Element	Symbol	At. No.	Density (g/cm^3)	Melting Pt. (°C)	Boiling Pt. (°C)
Tungsten	W	74	19.3	3410	5660
Uranium	U	92	18.95	1132	3818
Vanadium	V	23	6.11	1890	3380
Xenon	Xe	54	5.9 E-3 L: 3.52	−111.9	−107.1
Ytterbium	Yb	70	6.97	819	1194
Yttrium	Y	39	4.47	1522	3338
Zinc	Zn	30	7.13	419.58	907
Zirconium	Zr	40	6.51	1852	4377

TABLE A-2. Properties of the Elements: In Order of Atomic Number[a]

At. No.	Element	Symbol	Valence	Atomic Weight	Mass Number (in Order of Isotope Abundance)[b]
1	Hydrogen	H	1	1.00797	1, 2
2	Helium[c]	He	0	4.0026	4, 3
3	Lithium	Li	1	6.939	7, 6
4	Beryllium	Be	2	9.0122	9
5	Boron	B	3	10.811	11, 10
6	Carbon	C	±4, 2	12.01115	12, 13
7	Nitrogen	N	−3, 5, 2	14.0067	14, 15
8	Oxygen	O	−2	15.9994	16, 18, 17
9	Fluorine	F	−1	18.9984	19
10	Neon	Ne	0	20.183	20, 22, 21
11	Sodium	Na	1	22.9898	23
12	Magnesium	Mg	2	24.312	24, 26, 25
13	Aluminum	Al	3	26.9815	27
14	Silicon	Si	4	28.086	28, 29, 30
15	Phosphorus	P	5, ±3	30.9738	31
16	Sulfur	S	6, 4, −2	32.064	32, 34, 33, 36
17	Chlorine	Cl	±1, 7, 5	35.453	35, 37
18	Argon	Ar	0	39.948	40, 36, 38
19	Potassium	K	1	39.102	39, 41, (RA)40
20	Calcium	Ca	2	40.08	40, 44, 42, 48, 43, 46
21	Scandium	Sc	3	44.956	45
22	Titanium	Ti	4, 3	47.90	48, 46, 47, 49, 50
23	Vanadium	V	5, 4, 2	50.942	51, (RA)50
24	Chromium	Cr	6, 3, 2	51.996	52, 53, 50, 54
25	Manganese	Mn	7, 4, 2, 6, 3	54.9380	55
26	Iron	Fe	3, 2	55.847	56, 54, 57, 58
27	Cobalt	Co	3, 2	58.9332	59
28	Nickel	Ni	2, 3	58.71	58, 60, 62, 61, 64
29	Copper	Cu	2, 1	63.54	63, 65
30	Zinc	Zn	2	65.37	64, 66, 68, 67, 70
31	Gallium	Ga	3	69.72	69, 71
32	Germanium	Ge	4, 2	72.59	74, 72, 70, 76, 73
33	Arsenic	As	5, ±3	74.9216	75
34	Selenium	Se	6, 4, −2	78.96	80, 78, 82, 76, 77, 74
35	Bromine	Br	±1, 5	79.909	79, 81

36	Krypton	Kr	0	83.80	84, 86, 82, 76, 77, 74
37	Rubidium	Rb	1	85.47	85, (RA)87
38	Strontium	Sr	2	87.62	88, 86, 87, 84
39	Yttrium	Y	3	88.905	89
40	Zirconium	Zr	4	91.22	90, 94, 92, 91, 96
41	Niobium	Nb	5, 3	92.906	93
42	Molybdenum	Mo	6, 3, 5	95.94	98, 96, 92, 95, 100, 97, 94
43	Technetium	Te	7	98.9062	(RA:97, 98, 99)
44	Ruthenium	Ru	3, 4, 6, 8	101.07	102, 104, 101, 99, 100, 96, 98
45	Rhodium	Rh	3, 4	102.905	103
46	Palladium	Pd	2, 4	106.4	106, 109, 105, 110, 104, 102
47	Silver	Ag	1	107.868	107, 109
48	Cadmium	Cd	2	112.40	114, 112, 111, 110, 113, 116, 106, 108
49	Indium	In	3	114.82	115, 113
50	Tin	Sn	4, 2	118.69	120, 118, 116, 119, 117, 124, 122, 112, 114, 115
51	Antimony	Sb	3, 5	121.75	121, 123
52	Tellurium	Te	4, 6, −2	127.60	130, 128, 126, 125, 124, 122, 123, 120
53	Iodine	I	−1, 5, 7	126.9045	127
54	Xenon[c]	Xe	0	131.30	132, 129, 131, 134, 136, 130, 128, 124, 126
55	Cesium	Cs	1	132.9054	133
56	Barium	Ba	2	137.34	138, 137, 136, 135, 134, 130, 132
57	Lanthanum	La	3	138.9055	139, 138
58	Cerium[d]	Ce	3, 4	140.12	140, 142, 138, 136
59	Praseodymium[d]	Pr	3	140.9077	141
60	Neodymium[d]	Nd	3	144.24	142, 144, 146, 143, 145, 148, 150
61	Promethium[d]	Pm	3	(145)	(RA:145, 146, 147)
62	Samarium[d]	Sm	3	150.35	152, 154, 147, 149, 148, 150, 144
63	Europium[d]	Eu	3, 2	151.96	153, 151
64	Gadolinium[d]	Gd	3	157.25	158, 160, 156, 157, 155, 154, 152
65	Terbium[d]	Tb	3	158.9254	159
66	Dysprosium[d]	Dy	3	162.50	164, 162, 163, 161, 160, 158, 156
67	Holmium[d]	Ho	3	164.9304	165
68	Erbium[d]	Er	3	167.26	166, 168, 167, 170, 164, 162
69	Thulium[d]	Tm	3	168.9342	169
70	Ytterbium[d]	Yb	3, 2	173.04	174, 172, 173, 171, 176, 170, 168
71	Lutetium[d]	Lu	3	174.97	175, 176
72	Hafnium	Hf	4	178.49	180, 178, 177, 179, 176, 174
73	Tantalum	Ta	5	180.9479	181
74	Tungsten	W	6	183.85	184, 186, 182, 183, 180
75	Rhenium	Re	7, 4, −1	186.2	187, 185
76	Osmium	Os	4, 6, 8	190.2	192, 190, 189, 188, 187, 186, 184
77	Iridium	Ir	3, 4, 6	192.22	193, 191

TABLE A-2. (Continued)

At. No.	Element	Symbol	Valence	Atomic Weight	Mass Number (in Order of Isotope Abundance)[b]
78	Platinum	Pt	4, 2	195.09	195, 194, 196, 198, 192, 190
79	Gold	Au	3, 1	196.9665	197
80	Mercury	Hg	2, 1	200.59	202, 200, 199, 201, 198, 204, 196
81	Thallium	Tl	1, 3	204.37	205, 203
82	Lead	Pb	2, 4	207.19	208, 206, 207, 204
83	Bismuth	Bi	3, 5	208.9804	209
84	Polonium	Po	±2, 4, 0, 6	209	(RA:209)
85	Astatine	At	—	(210)	(RA:210, 211)
86	Radon[c]	Rn	0	(222)	(RA:222)
87	Francium	Fr	1	(223)	(RA:223, 212)
88	Radium	Ra	2	226.0254	(RA:226)
89	Actinium	Ac	3	(227)	(RA:228)
90	Thorium[e]	Th	4	232.0381	(RA)232
91	Protactinium[e]	Pa	5	321.0359	(RA:231)
92	Uranium[e]	U	6, 5, 4, 3	238.029	(RA)238, (RA)235, (RA)234
93	Neptunium[e]	Np	6, 5, 4, 3	237.0482	(RA:237)
94	Plutonium[e]	Pu	6, 5, 4, 3	(242)	(RA:244, 242, 239, 240)
95	Americium[e]	Am	3	(243)	(RA:243, 241)
96	Curium[e]	Cm	3	(247)	(RA:247)
97	Berkelium[e]	Bk	4, 3	(249)	(RA:247, 248, 249)
98	Californium[e]	Cf	3	(251)	(RA:251, 249)
99	Einsteinium[e]	Es	—	(252)	(RA:254, 252)
100	Fermium[e]	Fm	—	(257)	(RA:257)
101	Mendelevium[e]	Md	—	(258)	(RA:257)
102	Nobelium[e]	No	—	(255)	(RA:255, 253)
103	Lawrencium[e]	Lw	—	(256)	(RA:256, 258, 259)
104	Kurchatovium[f,g]	Ku	—	(261)	(RA:257, 259, 260)
105	Hahnium[f]	Ha	—	(262)	(RA:262, 261, 260)
106[f]		—	—	(263)	(RA:263)
107[f]		—	—	(264)	
109[f]		—	—	(266)	

[a]Selected and compiled from various sources; atomic weights are based on carbon-12.

[b]Stable or long-lived isotopes; "(RA: . . .)" = radioactive isotope(s) having longest half-life, or reasonably close half-life, in that order, where more than one is shown; "(RA)" = radioactive element or isotope whose abundance has been established.

[c]"Inert gas" ("noble gas").

[d]Lanthanide (series).

[e]Actinide (series).

[f]Tentatively identified; names not yet formally adopted.

[g]Alternative proposed name: Rutherfordium (Rf)

B

CONVERSION TABLES

TABLE B-1. Conversions of units of fluid-mechanical quantities

Density	kg/m^3
1 oz_m/in^3	1730
1 oz_m/gal	7.489
1 lb_m/ft^3	16.02
1 lb_m/in^3	2.768×10^4
1 lb_m/gal	119.8
Flow	m^3/s
1 gal/min	6.309×10^{-5}
1 ft^3/min	4.7195×10^{-4}
1 in^3/min	2.732×10^{-7}
	L/min
1 gal/min	3.7854
1 cm^3/min	1×10^{-3}
1 ft^3/min	28.317
	kg/s
1 lb_m/min	7.56×10^{-3}
1 lb_m/s	0.4536

Viscosity	SI unit
1 centipoise	10^{-3} N · s/m^2
1 centistoke	10^{-6} m^2/s
Pressure	**SI unit**
atmosphere	101.3 kPa
bar	100 kPa (10^5 Pa)
in Hg (0°C)	3.3864 kPa
kg_f/cm^2	98.07 kPa
kg_f/m^2	9.80665 Pa
mbar	100 Pa
mm Hg (0°C)	133.32 Pa
torr	133.32 Pa

Pressure Unit Conversions, psi to kPa and MPA

psi	kPa
(1.000	6.895)
1.00	6.9
2.00	13.8
3.00	20.7
4.00	27.6
5.00	34.5
6.00	41.4
7.00	48.3
8.00	55.2
9.00	62.1
10.0	69
20.0	138
30.0	207
40.0	276
50.0	345
60.0	414
70.0	483
80.0	552
90.0	621
100	690

psi	MPa
100	0.69
145	1.00
200	1.38
300	2.07
400	2.76
500	3.45
600	4.14
700	4.83
800	5.52
900	6.21
1000	6.90
1450	10.00
2000	13.79
5000	34.47
10 000	68.95
20 000	137.9

TABLE B-2. Temperature scale conversion

Celsius (°C)	Fahrenheit (°F)	Kelvin (K)
Absolute Zero −273.15	−459.67	0
−272.0	−457.6	1.1
−270.0	−454.0	3.1
−268.0	−450.4	5.1
−266.0	−446.8	7.1
−265.0	−445.0	8.2
−264.0	−443.2	9.1
−262.0	−439.6	11.1
−260.0	−436.0	13.1
−258.0	−432.4	15.1
−256.0	−428.8	17.1
−255.0	−427.0	18.1
−254.0	−425.2	19.1
−252.0	−421.6	21.1
−250.0	−418.0	23.1
−248.0	−414.4	25.1
−245.0	−409.0	28.1
−244.0	−407.2	29.1
−242.0	−403.6	31.1
−240.0	−400.0	33.1

Celsius (°C)	Fahrenheit (°F)	Kelvin (K)
−238.0	−396.4	35.1
−236.0	−392.8	37.1
−235.0	−391.0	38.1
−234.0	−389.2	39.1
−232.0	−385.6	41.1
−230.0	−382.0	43.1
−228.0	−378.4	45.1
−226.0	−374.8	47.1
−225.0	−373.0	48.1
−224.0	−371.2	49.1
−222.0	−367.6	51.1
−220.0	−364.0	53.1
−218.0	−360.4	55.1
−216.0	−356.8	57.1
−215.0	−355.0	58.1
−214.0	−353.2	59.1
−212.0	−349.6	61.1
−210.0	−346.0	63.1
−208.0	−342.4	65.1
−206.0	−338.8	67.1
−205.0	−337.0	68.1

TABLE B-2. (Continued)

Celsius (°C)	Fahrenheit (°F)	Kelvin (K)	Celsius (°C)	Fahrenheit (°F)	Kelvin (K)
−204.0	−335.2	69.1	−88.0	−126.4	185.1
−202.0	−331.6	71.1	−84.0	−119.2	189.1
−200.0	−328.0	73.1	−80.0	−112.0	193.1
−198.0	−324.4	75.1	−78.9	−110.0	194.3
−196.0	−320.8	77.1	−76.0	−104.8	197.1
−195.0	−319.0	78.1	−75.0	−103.0	198.1
−194.0	−317.2	79.1	−72.0	−97.6	201.1
−192.0	−313.6	81.1	−68.0	−90.4	205.1
−190.0	−310.0	83.1	−64.0	−83.2	209.1
−188.0	−306.4	85.1	−60.0	−76.0	213.1
−186.0	−302.8	87.1	−56.0	−68.8	217.1
−185.0	−301.0	88.1	−55.0	−67.0	218.1
−184.0	−299.2	89.1	−52.0	−61.6	221.1
−182.2	−296.0	90.9	−48.0	−54.4	225.1
−180.0	−292.0	93.1	−44.0	−47.2	229.1
−178.0	−288.4	95.1	−43.3	−46.0	229.8
−176.0	−284.8	97.1	−40.0	−40.0	233.1
−175.0	−283.0	98.1	−38.9	−38.0	234.3
−174.0	−281.2	99.1	−36.0	−32.8	237.1
−172.0	−277.6	101.1	−35.0	−31.0	238.1
−170.0	−274.0	103.1	−32.0	−25.6	241.1
−168.0	−270.4	105.1	−31.1	−24.0	242.0
−166.0	−266.8	107.1	−28.0	−18.4	245.1
−165.0	−265.0	108.1	−27.2	−17.0	245.9
−164.0	−263.2	109.1	−24.0	−11.2	249.1
−162.0	−259.6	111.1	−23.3	−10.0	249.8
−160.0	−256.0	113.1	−20.0	−4.0	253.1
−158.0	−252.4	115.1	−16.0	3.2	257.1
−156.0	−248.8	117.1	−15.6	4.0	257.6
−155.0	−247.0	118.1	−12.0	10.4	261.1
−154.0	−245.2	119.1	−11.7	11.0	261.5
−152.0	−241.6	121.1	−8.0	17.6	265.1
−150.0	−238.0	123.1	−7.8	18.0	265.4
−148.0	−234.4	125.1	−4.0	24.8	269.1
−144.0	−227.2	129.1	−3.9	25.0	269.3
−140.0	−220.0	133.1	0.0	32.0	273.15
−136.0	−212.8	137.1	4.0	39.2	277.1
−135.0	−211.0	138.1	4.4	40.0	277.6
−132.0	−205.6	141.1	8.0	46.4	281.1
−128.0	−198.4	145.1	8.3	47.0	281.5
−124.0	−191.2	149.1	12.0	53.6	285.1
−120.0	−184.0	153.1	12.2	54.0	285.4
−116.0	−176.8	157.1	16.0	60.8	289.1
−115.0	−175.0	158.1	16.1	61.0	289.3
−112.0	−169.6	161.1	20.0	68.0	293.1
−108.0	−162.4	165.1	24.0	75.2	297.1
−104.0	−155.2	169.1	24.4	76.0	297.6
−100.0	−148.0	173.1	28.0	82.4	301.1
−96.0	−140.8	177.1	28.3	83.0	301.5
−95.0	−139.0	178.1	32.0	89.6	305.1
−92.0	−133.6	181.1	32.2	90.0	305.4

TABLE B-2. (Continued)

Celsius (°C)	Fahrenheit (°F)	Kelvin (K)	Celsius (°C)	Fahrenheit (°F)	Kelvin (K)
36.0	96.8	309.1	232.0	449.6	505.1
36.1	97.0	309.3	236.0	456.8	509.1
40.0	104.0	313.1	240.0	464.0	513.1
44.0	111.2	317.1	244.0	471.2	517.1
44.4	112.0	317.6	248.0	478.4	521.1
48.0	118.4	321.1	252.0	485.6	525.1
52.0	125.6	325.1	256.0	492.8	529.1
56.0	132.8	329.1	260.0	500.0	533.1
60.0	140.0	333.1	264.0	507.2	537.1
64.0	147.2	337.1	268.0	514.4	541.1
68.0	154.4	341.1	272.2	522.0	545.4
72.0	161.6	345.1	276.0	528.8	549.1
76.0	168.8	349.1	280.0	536.0	553.1
80.0	176.0	353.1	284.0	543.2	557.1
84.0	183.2	357.1	288.0	550.4	561.1
88.0	190.4	361.1	292.0	557.6	565.1
92.0	197.6	365.1	296.0	564.8	569.1
96.0	204.8	369.1	300.0	572.0	573.1
104.0	219.2	377.1	304.0	579.2	577.1
108.0	226.4	381.1	308.0	586.4	581.1
112.0	233.6	385.1	312.0	593.6	585.1
116.0	240.8	389.1	316.0	600.8	589.1
120.0	248.0	393.1	320.0	608.0	593.1
124.0	255.2	397.1	324.0	615.2	597.1
128.0	262.4	401.1	328.0	622.4	601.1
132.0	269.6	405.1	332.0	629.6	605.1
136.0	276.8	409.1	336.0	636.8	609.1
140.0	284.0	413.1	340.0	644.0	613.1
144.0	291.2	417.1	344.0	651.2	617.1
148.0	298.4	421.1	348.0	658.4	621.1
152.0	305.6	425.1	352.0	665.6	625.1
156.0	312.8	429.1	356.0	672.8	629.1
160.0	320.0	433.1	360.0	680.0	633.1
164.0	327.2	437.1	364.0	687.2	637.1
168.0	334.4	441.1	368.0	694.4	641.1
172.0	341.6	445.1	372.0	701.6	645.1
176.0	348.8	449.1	376.0	708.8	649.1
180.0	356.0	453.1	380.0	716.0	653.1
184.0	363.2	457.1	384.0	723.2	657.1
188.0	370.4	461.1	388.0	730.4	661.1
192.0	377.6	465.1	392.0	737.6	665.1
196.0	384.8	469.1	396.0	744.8	669.1
200.0	392.0	473.1	400.0	752.0	673.1
204.0	399.2	477.1	404.0	759.2	677.1
208.0	406.4	481.1	408.0	766.4	681.1
212.0	413.6	485.1	412.0	773.6	685.1
216.0	420.8	489.1	416.0	780.8	689.1
220.0	428.0	493.1	420.0	788.0	693.1
224.0	435.2	497.1	424.0	795.2	697.1
228.0	442.4	501.1	428.0	802.4	701.1

TABLE B-2. (Continued)

Celsius (°C)	Fahrenheit (°F)	Kelvin (K)
432.0	809.6	705.1
436.0	816.8	709.1
440.0	824.0	713.1
444.0	831.2	717.1
445.0	833.0	718.1
448.0	838.4	721.1
452.0	845.6	725.1
456.0	852.8	729.1
460.0	860.0	733.1
464.0	867.2	737.1
468.0	874.4	741.1
472.0	881.6	745.1
476.0	888.8	749.1
480.0	896.0	753.1
484.0	903.2	757.1
488.0	910.4	761.1
492.0	917.6	765.1
496.0	924.8	769.1
500.0	932.0	773.1
504.0	939.2	777.1
508.0	946.4	781.1
512.0	953.6	785.1
516.0	960.8	789.1
520.0	968.0	793.1
524.0	975.2	797.1
528.0	982.4	801.1
532.0	989.6	805.1
536.0	996.8	809.1
540.0	1004.	813.1
544.0	1011.	817.1
548.0	1018.	821.1
552.0	1026.	825.1
556.0	1033.	829.1
560.0	1040.	833.1
564.0	1047.	837.1
568.0	1054.	841.1
572.2	1062.	845.1
576.0	1063.	849.1
580.0	1076.	853.1
584.0	1083.	857.1
588.0	1090.	861.1
592.0	1098.	865.1
596.0	1105.	869.1
600.0	1112.	873.1
604.0	1119.	877.1
608.0	1126.	881.1
612.0	1134.	885.1
616.0	1141.	889.1
620.0	1148.	893.1
624.0	1155.	897.1
628.0	1162.	901.1
632.2	1170.	905.4
636.0	1177.	909.1
640.0	1184.	913.1
644.0	1191.	917.1
648.0	1198.	921.1
652.0	1206.	925.1
656.0	1213.	929.1
660.0	1220.	933.1
664.0	1227.	937.1
668.0	1234.	941.1
672.0	1242.	945.1
676.0	1249.	949.1
680.0	1256.	953.1
684.0	1263.	957.1
688.0	1270.	961.1
692.0	1278.	965.1
696.0	1285.	969.1
700.0	1292.	973.1
704.0	1299.	977.1
708.0	1306.	981.1
712.0	1314.	985.1
716.0	1321.	989.1
720.0	1328.	993.1
724.0	1335.	997.1
728.0	1342.	1001.
732.0	1350.	1005.
736.0	1357.	1009.
740.0	1364.	1013.
744.0	1371.	1017.
748.0	1378.	1021.
752.0	1386.	1025.
756.0	1393.	1029.
760.0	1400.	1033.
764.0	1407.	1037.
768.0	1414.	1041.
772.0	1422.	1045.
776.0	1429.	1049.
780.0	1436.	1053.
784.0	1443.	1057.
784.4	1444.	1058.
788.0	1450.	1061.
788.3	1451.	1062.
792.0	1458.	1065.
796.0	1465.	1069.
800.0	1472.	1073.
804.0	1479.	1077.
808.0	1486.	1081.
812.0	1494.	1085.
816.0	1501.	1089.

TABLE B-2. (Continued)

Celsius (°C)	Fahrenheit (°F)	Kelvin (K)	Celsius (°C)	Fahrenheit (°F)	Kelvin (K)
820.0	1508.	1093.	1020.	1868.	1293.
824.0	1515.	1097.	1024.	1875.	1297.
828.0	1522.	1101.	1028.	1882.	1301.
832.0	1530.	1105.	1032.	1890.	1305.
836.0	1537.	1109.	1036.	1897.	1309.
840.0	1544.	1113.	1040.	1904.	1313.
844.0	1551.	1117.	1044.	1911.	1317.
848.0	1558.	1121.	1048.	1918.	1321.
852.0	1566.	1125.	1052.	1926.	1325.
856.0	1573.	1129.	1056.	1933.	1329.
860.0	1580.	1133.	1060.	1940.	1333.
864.0	1587.	1137.	1064.	1947.	1337.
868.0	1594.	1141.	1068.	1954.	1341.
872.0	1602.	1145.	1072.	1962.	1345.
876.0	1609.	1149.	1076.	1969.	1349.
880.0	1616.	1153.	1080.	1976.	1353.
884.0	1623.	1157.	1084.	1983.	1357.
888.0	1630.	1161.	1088.	1990.	1361.
892.0	1638.	1165.	1092.	1998.	1365.
896.0	1645.	1169.	1096.	2005.	1369.
900.0	1652.	1173.	1100.	2012.	1373.
904.0	1659.	1177.	1110.	2030.	1383.
908.0	1666.	1181.	1120.	2048.	1393.
912.0	1674.	1185.	1130.	2066.	1403.
916.0	1681.	1189.	1140.	2084.	1413.
920.0	1688.	1193.	1150.	2102.	1423.
924.0	1695.	1197.	1160.	2120.	1433.
928.0	1702.	1201.	1170.	2138.	1443.
932.0	1710.	1205.	1180.	2156.	1453.
936.0	1717.	1209.	1190.	2174.	1463.
940.0	1724.	1213.	1200.	2192.	1473.
944.0	1731.	1217.	1210.	2210.	1483.
948.0	1738.	1221.	1220.	2228.	1493.
952.0	1746.	1225.	1230.	2246.	1503.
956.0	1753.	1229.	1240.	2264.	1513.
960.0	1760.	1233.	1250.	2282.	1523.
964.0	1767.	1237.	1260.	2300.	1533.
968.0	1774.	1241.	1270.	2318.	1543.
972.2	1782.	1245.	1280.	2336.	1553.
976.0	1789.	1249.	1290.	2354.	1563.
980.0	1796.	1253.	1300.	2372.	1573.
984.0	1803.	1257.	1310.	2390.	1583.
988.0	1810.	1261.	1320.	2408.	1593.
992.0	1818.	1265.	1330.	2426.	1603.
996.0	1825.	1269.	1340.	2444.	1613.
1000.	1832.	1273.	1350.	2462.	1623.
1004.	1839.	1277.	1360.	2480.	1633.
1008.	1846.	1281.	1370.	2498.	1643.
1012.	1854.	1285.	1380.	2516.	1653.
1016.	1861.	1289.	1390.	2534.	1663.

TABLE B-2. (Continued)

Celsius (°C)	Fahrenheit (°F)	Kelvin (K)	Celsius (°C)	Fahrenheit (°F)	Kelvin (K)
1400.	2552.	1673.	1920.	3488.	2193.
1420.	2588.	1693.	1930.	3506.	2203.
1430.	2606.	1703.	1940.	3524.	2213.
1440.	2624.	1713.	1950.	3542.	2223.
1450.	2642.	1723.	1960.	3560.	2233.
1460.	2660.	1733.	1970.	3578.	2243.
1470.	2678.	1743.	1980.	3596.	2253.
1480.	2696.	1753.	1990.	3614.	2263.
1490.	2714.	1763.	2000.	3632.	2273.
1500.	2732.	1773.	2010.	3650.	2283.
1510.	2750.	1783.	2020.	3668.	2293.
1520.	2768.	1793.	2030.	3686.	2303.
1530.	2786.	1803.	2040.	3704.	2313.
1540.	2804.	1813.	2050.	3722.	2323.
1550.	2822.	1823.	2060.	3740.	2333.
1560.	2840.	1833.	2070.	3758.	2343.
1570.	2858.	1843.	2080.	3776.	2353.
1580.	2876.	1853.	2090.	3794.	2363.
1590.	2894.	1863.	2100.	3812.	2373.
1600.	2912.	1873.	2110.	3830.	2383.
1610.	2930.	1883.	2120.	3848.	2393.
1620.	2948.	1893.	2130.	3866.	2403.
1630.	2966.	1903.	2140.	3884.	2413.
1640.	2984.	1913.	2150.	3902.	2423.
1650.	3002.	1923.	2160.	3920.	2433.
1670.	3038.	1943.	2170.	3938.	2443.
1680.	3056.	1953.	2180.	3956.	2453.
1690.	3074.	1963.	2190.	3974.	2463.
1700.	3092.	1973.	2200.	3992.	2473.
1710.	3110.	1983.	2210.	4010.	2483.
1720.	3128.	1993.	2220.	4028.	2493.
1730.	3146.	2003.	2230.	4046.	2503.
1740.	3164.	2013.	2240.	4064.	2513.
1750.	3182.	2023.	2250.	4082.	2523.
1760.	3200.	2033.	2260.	4100.	2533.
1770.	3218.	2043.	2270.	4118.	2543.
1780.	3236.	2053.	2280.	4136.	2553.
1790.	3254.	2063.	2290.	4154.	2563.
1800.	3272.	2073.	2300.	4172.	2573.
1810.	3290.	2083.	2310.	4190.	2583.
1820.	3308.	2093.	2320.	4208.	2593.
1830.	3326.	2103.	2330.	4226.	2603.
1840.	3344.	2113.	2340.	4244.	2613.
1850.	3362.	2123.	2350.	4262.	2623.
1860.	3380.	2133.	2360.	4280.	2633.
1870.	3398.	2143.	2370.	4298.	2643.
1880.	3416.	2153.	2380.	4316.	2653.
1890.	3434.	2163.	2390.	4334.	2663.
1900.	3452.	2173.	2400.	4352.	2673.
1910.	3470.	2183.	2420.	4388.	2693.

TABLE B-2. (Continued)

Celsius (°C)	Fahrenheit (°F)	Kelvin (K)
2440.	4424.	2713.
2460.	4460.	2733.
2480.	4496.	2753.
2500.	4532.	2773.
2520.	4568.	2793.
2540.	4604.	2813.
2560.	4640.	2833.
2580.	4676.	2853.
2600.	4712.	2873.
2620.	4748.	2893.
2640.	4784.	2913.
2660.	4820.	2933.
2680.	4856.	2953.
2700.	4892.	2973.
2720.	4928.	2993.
2740.	4964.	3013.
2760.	5000.	3033.
2780.	5036.	3053.
2800.	5072.	3073.
2820.	5108.	3093.
2840.	5144.	3113.
2860.	5180.	3133.
2880.	5216.	3153.
2900.	5252.	3173.
2920.	5288.	3193.
2940.	5324.	3213.
2960.	5360.	3233.
2980.	5396.	3253.
3000.	5432.	3273.
3020.	5468.	3293.
3040.	5504.	3313.
3060.	5540.	3333.
3080.	5576.	3353.
3100.	5612.	3373.
3120.	5648.	3393.
3140.	5684.	3413.
3160.	5720.	3433.
3180.	5756.	3453.
3200.	5792.	3473.
3220.	5828.	3493.
3240.	5864.	3513.
3260.	5900.	3533.
3280.	5936.	3553.
3300.	5972.	3573.
3320.	6008.	3593.
3340.	6044.	3613.
3360.	6080.	3633.
3380.	6116.	3653.
3400.	6152.	3673.
3420.	6188.	3693.
3440.	6224.	3713.
3460.	6260.	3733.
3480.	6296.	3753.
3500.	6332.	3773.

C

INSTRUMENT MATERIAL COMPATIBILITY

Table C-1. Recommendations for Sensor Materials in Contact with Measured Fluids

Substance	Conditions	Recommended Metal
Acetate solvents	Crude or pure	Monel or nickel
Acetic acid	10%, 70°F	304 stainless steel
	50%, 70°F	304 Stainless Steel
	50%, 212°F	316 stainless steel
	99%, 70°F	430 stainless steel
	99%, 212°F	430 stainless steel
Acetic anhydride		Monel
Acetone	212°F	304 stainless steel
Acetylene		304, Monel, nickel
Alcohol ethyl	70°F	304 stainless steel
	212°F	304 stainless steel
Alcohol methyl	70°F	304 stainless steel
	212°F	304 stainless steel
Aluminum	Molten	Cast iron
Aluminum acetate	Saturated	304 stainless steel
Aluminum sulfate	10%, 70°F	304 stainless steel
	Saturated 70°F	304 stainless steel
	10%, 212°F	316 stainless steel
	Saturated, 212°F	316 stainless steel
Ammonia	All concentrations, 70°F	304 stainless steel
Ammonium chloride	All concentrations, 212°F	316 stainless steel

Table C-1. (Continued)

Substance	Conditions	Recommended Metal
Ammonium nitrate	All concentrations, 70°F	304 stainless steel
	All concentrations, 212°F	304 stainless steel
Ammonium sulfate	5%, 70°F	304 stainless steel
	10%, 212°F	316 stainless steel
	Saturated, 212°F	316 stainless steel
Aniline	All concentrations, 70°F	304 stainless steel
Amylacetate		Monel
Asphalt		Steel (C1018), phosphor bronze, Monel, nickel
Barium carbonate	70°F	304 stainless steel
Barium chloride	5%, 70°F	Monel
	Saturated, 70°F	Monel
	Aqueous, hot	316 stainless steel
Barium hydroxide		Steel (C1018)
Barium sulfate		Nichrome
Benzaldehyde		Steel (C1018)
Benzene	70°F	304 stainless steel
Benzine		Steel (C1018), Monel, Inconel
Benzol	Hot	304 stainless steel
Boracic acid	5%, hot or cold	304 stainless steel
Bromine	70°F	Tantalum
Butadiene		Brass, 304 stainless steel
Butane	70°F	304 stainless steel
Butylacetate		Monel
Butyl alcohol		Copper
Butylenes		Steel (C1018), phosphor bronze
Butyric acid	5%, 70°F	304 stainless steel
	5%, 150°F	304 stainless steel
Calcium bisulfite	70°F	316 stainless steel
Calcium chlorate	Dilute, 70°F	304 stainless steel
	Dilute, 150°F	304 stainless steel
Calcium hydroxide	10%, 212°F	304 stainless steel
	20%, 212°F	304 stainless steel
	50%, 212°F	317 stainless steel
Carbolic acid	All, 212°F	316 stainless steel
Carbon dioxide	Dry	Steel (C1018), Monel
Carbon dioxide	Wet	Aluminum, Monel, nickel
Carbon tetrachloride	10%, 70°F	Monel
Chlorex caustic		316, 317 stainless steel
Chlorine gas	Dry, 70°F	317 stainless steel
	Moist, 70°F	Hastelloy C
	Moist, 212°F	Hastelloy C
Chromic acid	5%, 70°F	304 stainless steel
	10%, 212°F	316 stainless steel
	50%, 212°F	316 stainless steel
Citric acid	15%, 70°F	304 stainless steel
	15%, 212°F	316 stainless steel
	Concentrated, 212°F	317 stainless steel
Coal tar	Hot	304 stainless steel
Coke oven gas		Aluminum
Copper nitrate		304, 316 stainless steel
Copper sulfate		304, 316 stainless steel

Core oils		316 stainless steel
Cottonseed oil		Steel (C1018), Monel, nickel
Creosols		304 stainless steel
Creosote crude		Steel (C1018), Monel, nickel
Cyanogen gas		304 stainless steel
Dowtherm		Steel (C1018)
Epsom salt	Hot and cold	304 stainless steel
Ether	70°F	304 stainless steel
Ethyl acetate		Monel
Ethyl chloride	70°F	304 stainless steel
Ethylene glycol		Steel (C1018)
Ethyl sulfate	70°F	Monel
Ferric chloride	1%, 70°F	316 stainless steel
	5%, 70°F	Tantalum
	5%, boiling	Tantalum
Ferric sulfate	5%, 70°F	304 stainless steel
Ferrous sulfate	Dilute, 70°F	304 stainless steel
Formaldehyde		304 stainless steel
Freon		Steel (C1018)
Formic acid	5%, 70°F	316 stainless steel
	5%, 150°F	316 stainless steel
Gallic acid	5%, 70°F	Monel
	5%, 150°F	Monel
Gasoline	70°F	304 stainless steel
Glucose	70°F	304 stainless steel
Glycerine	70°F	304 stainless steel
Glycerol		304 stainless steel
Heat treating		446 stainless steel
Hydrobromic acid	48%, 212°F	Hastelloy B
Hydrochloric acid	1%, 70°F	Hastelloy C
	1%, 212°F	Hastelloy B
	5%, 70°F	Hastelloy C
	5%, 212°F	Hastelloy B
	25%, 70°F	Hastelloy B
	25%, 212°F	Hastelloy B
Hydrocyanic acid		316 stainless steel
Hydrofluoric acid		Hastelloy C
Hydrogen peroxide	70°F	316 stainless steel
	212°F	316 stainless steel
Hydrogen sulfide	Wet and dry	316 stainless steel
Iodine	70°F	Tantalum
Kerosene	70°F	304 stainless steel
Lactic acid	5%, 70°F	304 stainless steel
	5%, 150°F	316 stainless steel
	10%, 212°F	Tantalum
Lacquer	70°F	316 stainless steel
Latex		Steel (C1018)
Lime sulfur		Steel (C1018), 304 stainless steel, Monel
Linseed oil	70°F	304 stainless steel
Magnesium chloride	5%, 70°F	Monel
	5%, 212°F	Nickel
Magnesium sulfate	Cold and hot	Monel
Malic acid	Cold and hot	316 stainless steel
Mercury		Steel (C1018), 304 stainless steel, Monel
Methane	70°F	Steel (1020)
Milk		304 stainless steel, Nickel

Table C-1. (Continued)

Substance	Conditions	Recommended Metal
Mixed acids (sulfuric and nitric —all temp. and %)		Carpenter 20
Molasses		Steel (C1018), 304 stainless steel, Monel, nickel
Muriatic acid	70°F	Tantalum
Naptha	70°F	304 stainless steel
Natural gas	70°F	304 stainless steel
Neon	70°F	304 stainless steel
Nickel chloride	70°F	304 stainless steel
Nickel sulfate	Hot and cold	304 stainless steel
Nitric acid	5%, 70°F	304 stainless steel
	20%, 70°F	304 stainless steel
	50%, 70°F	304 stainless steel
	50%, 212°F	304 stainless steel
	65%, 212°F	316 stainless steel
	Concentrated, 70°F	304 stainless steel
	Concentrated, 212°F	Tantalum
Nitrobenzene	70°F	304 stainless steel
Nitrous acid		304 stainless steel
Oleic acid	70°F	316 stainless steel
Oleum	70°F	316 stainless steel
Oxolic acid	5%, Hot and cold	304 stainless steel
	10%, 212°F	Monel
Oxygen	70°F	Steel (C1018)
Oxygen	Liquid	304 stainless steel
Palmitic acid		316 stainless steel
Petroleum ether		304 stainless steel
Phenol		304 stainless steel
Pentane		304 stainless steel
Phosphoric acid	1%, 70°F	304 stainless steel
	5%, 70°F	304 stainless steel
	10%, 70°F	316 stainless steel
	10%, 212°F	Hastelloy C
	30%, 70°F	Hastelloy B
	30%, 212°F	Hastelloy B
	85%, 70°F	Hastelloy B
	85%, 212°F	Hastelloy B
Picric acid	70°F	304 stainless steel
Potassium bromide	70°F	316 stainless steel
Potassium carbonate	1%, 70°F	304 stainless steel
Potassium chlorate	70°F	304 stainless steel
Potassium chloride	5%, 70°F	304 stainless steel
	5%, 212°F	304 stainless steel
Potassium hydroxide	5%, 70°F	304 stainless steel
	25%, 212°F	304 stainless steel
	50%, 212°F	316 stainless steel
Potassium nitrate	5%, 70°F	304 stainless steel
	5%, 212°F	304 stainless steel
Potassium Permanganate	5%, 70°F	304 stainless steel
Potassium sulfate	5%, 70°F	304 stainless steel
	5%, 212°F	304 stainless steel

Potassium sulfide	70°F	304 stainless steel
Propane		304 stainless steel
Pyrogallic acid		304 stainless steel
Quinine bisulfate	Dry	316 stainless steel
Quinine sulfate	Dry	304 stainless steel
Resin		304 stainless steel
Rosin	Molten	304 stainless steel
Seawater		Monel
Salammoniac		Monel
Salicylic acid		Nickel
Shellac		304 stainless steel
Soap	70°F	304 stainless steel
Sodium bicarbonate	All concentrations, 70°F	304 stainless steel
	5%, 150°F	304 stainless steel
Sodium bisulfate		Monel
Sodium carbonate	5%, 70°F	304 stainless steel
	5%, 150°F	304 stainless steel
Sodium chloride	5%, 70°F	316 stainless steel
	5%, 150°F	316 stainless steel
	Saturated, 70°F	316 stainless steel
	Saturated, 212°F	316 stainless steel
Sodium fluoride	5%, 70°F	Monel
Sodium hydroxide		304 stainless steel
Sodium hypochlorite	5% still	316 stainless steel
Sodium nitrate	Fused	317 stainless steel
Sodium peroxide		304 stainless steel
Sodium phosphate		Steel (C1018)
Sodium silicate		Steel (C1018)
Sodium sulfate	70°F	304 stainless steel
Sodium sulfide	70°F	316 stainless steel
Sodium sulfite	150°F	304 stainless steel
Steam		304 stainless steel
Stearic acid		304 stainless steel
Sulphur dioxide	Moist gas, 70°F	316 stainless steel
	Gas, 575°F	304 stainless steel
Sulphur	Dry, molten	304 stainless steel
	Wet	316 stainless steel
Sulphuric acid	5%, 70°F	Carp. 20, Hastelloy B
	5%, 212°F	Carp. 20, Hastelloy B
	10%, 70°F	Carp. 20, Hastelloy B
	10%, 212°F	Carp. 20, Hastelloy B
	50%, 70°F	Carp. 20, Hastelloy B
	50%, 212°F	Carp. 20, Hastelloy B
	90%, 70°F	Carp. 20, Hastelloy B
	90%, 212°F	Hastelloy D
Tannic acid	70°F	304 stainless steel
Tar		Steel (C1018), 304 stainless steel, Monel, nickel
Tartaric acid	70°F	304 stainless steel
	150°F	316 stainless steel
Tin	Molten	Cast iron
Toluene		Aluminum, phosphor bronze, Monel
Trichloro ethylene		Steel (C1018)
Turpentine		304 stainless steel
Varnish		304 stainless steel
Vegetable oils		Steel (C1018), 304 stainless steel, Monel
Vinegar		304 stainless steel

Table C-1. (Continued)

Substance	Conditions	Recommended Metal
Water	Fresh	Copper, steel (C1018), Monel
Water	Salt	Aluminum, brass, Monel
Whiskey, wine		304 stainless steel, nickel
Xylene		Copper
Zinc	Molten	Cast iron
Zinc chloride		Monel
Zinc sulfate	5%, 70°F	304 stainless steel
	Saturated, 70°F	304 stainless steel
	25%, 212°F	304 stainless steel

Courtesy of Omega Engineering, Inc., an Omega Group Co.

Notes:

1. This table considers only metallic materials used for portions of instruments (sensing portions or sensors) that come in contact with the sample (measured fluid, process liquid, or gas); sensing portion materials such as glass, ceramic, or plastic are not included.
2. Considerations of compatibility include contamination, electrolysis, and catalytic reactions.
3. Many of the recommended metals are proprietary alloys used primarily in the U.S.A.; in most cases, equivalent alloys exist in other countries.

INDEX

Q

R

S